MW00587074

CULL OF THE WILD

CULL OF THE WILD

Killing in the Name of Conservation

Hugh Warwick

BLOOMSBURY WILDLIFE
LONDON • OXFORD • NEW YORK • NEW DELHI • SYDNEY

BLOOMSBURY WILDLIFE
Bloomsbury Publishing Plc
50 Bedford Square, London, WC1B 3DP, UK
29 Earlsfort Terrace, Dublin 2, Ireland

BLOOMSBURY, BLOOMSBURY WILDLIFE and the Diana logo are trademarks of Bloomsbury Publishing Plc

First published in the United Kingdom 2024

For legal purposes the Acknowledgements on p. 291 constitute an extension of this copyright page

A catalogue record for this book is available from the British Library.

Library of Congress Cataloguing-in-Publication data has been applied for.

ISBN: Hardback: 978-1-3994-0374-0; ePub: 978-1-3994-0372-6;
ePDF: 978-1-3994-0373-3

2 4 6 8 10 9 7 5 3 1

Typeset in Bembo Std by Deanta Global Publishing Services, Chennai, India
Printed and bound in Great Britain by CPI Group (UK) Ltd, Croydon CR0 4YY

To Pat and Mary Morris – thank you for your kindness

Contents

Introduction

Every year millions of animals are killed in the name of conservation.

The animals that are killed can be native or introduced, they can be feral individuals or populations of domestic animals. The means by which they are killed is often ungoverned, and can be cruel. This practice receives very little attention. It is often referred to as culling, softer in tone than killing. We tend not to think too much about it for the obvious reason that it is really rather disturbing. And conservationists tend not to talk too much about it for the obvious reason that they rely heavily on our support.

But conservationists are the good guys, aren't they? They are the ones trying to save the planet, to rescue species from extinction. We donate billions of pounds to enable them to take a stand. And I support the work of conservationists through word and deed.

I believe we deserve an honest conversation about conservation. To do that we need to establish one very important point. Conservation, wildlife management, and the ecology that underpins them both, is *really* complicated. Add to this one more variable: people with differing perspectives. Now, it becomes close to impossible to solve the very real problems with which we are confronted. Complicated problems rarely have simple solutions – there is an old saying that anyone who gives you a simple answer to a complicated problem is either a liar or a fool.

The reason I believe a light should be shone into this less-examined corner of conservation is not to berate the hard-working wildlife managers, though I anticipate many readers will feel an initial surge of outrage as reality is exposed. No, I am hoping that this will encourage an acknowledgement that this work is really complicated – and that maybe we should all start to treat the subject of ecology with a little more respect.

Animals are being killed because when they are in (what we consider to be) the wrong place at the wrong time, they can have a seriously detrimental impact on the ability of other species to thrive. So serious is this impact that the United Nations Convention on Biological Diversity lists it in the top five most significant threats to biodiversity and bioabundance, alongside more commonly recognised problems:

- Habitat loss and degradation
- Climate change
- Excessive nutrient load and other forms of pollution
- Over-exploitation and unsustainable use
- Invasive alien species

The first four of these threats are things that we need to *stop* doing. With alien species, we can actively do some good.

A great many alien species are plants, and there are even more plants killed each year than animals. But I am going to focus in this book on animals, because that is where the most heated debates are taking place.

I want this book to be honest. So, to start how I mean to go on, let us consider one very important word:

Prejudice: noun /ˈprɛdʒʊdɪs/ – preconceived opinion that is not based on reason or actual experience.

We all have them; our own private and not-so-private opinions that affect our judgement. In an ideal world we could solve problems by objectively looking at the evidence and seeing the way forward. But ecological problems are multi-layered.

Not just the web of life with its uncountable interactions – rocks that influence soils that influence microbes that influence plants that influence insects that influence amphibians and reptiles, birds and mammals. No, not just that, but also us – humans, with all of our prejudices that help contradict and confound what the evidence might suggest.

My prejudices are clear, and seem to be innate. My earliest memories had me tied to animals more than people. It was mostly mammals; I never really got the bird bug. Perhaps it gives a glimpse at the nature-nurture argument – my adopted parents were not particularly interested in the natural world, but my biological mother, who I met when I was in my mid thirties, has always had a deep love for nature, is an avid gardener and has carried the nickname 'Badger' since her schooldays.

I have already used the 'L' word... many people feel shy about saying they love the wildlife which supports us all; that it will leave them open to being dismissed as cranks. I am reminded that in mechanics, cranks are what drive revolutions!

The honesty is going to be, I imagine, at times uncomfortable or even painful. As we examine our relationship with nature, I fear I will have my prejudices challenged, and that is never pleasant.

I feel very strongly about animals – and about nature. I find it hits me hard in the heart when I see and hear of cruelty, and I can find myself despairing at the state of humanity. There are things about which I have feelings so strong that they interfere with my ability to be reasonable. People who take pleasure from causing pain and death to animals revolt me. I stopped eating meat more than 35 years ago, spent 10 years as a strict vegan and am currently what I describe as a 'vague-an' ... vaguely vegan. Soya milk in tea, oat milk on muesli, but cake ... well that is another category, and as I do hundreds of talks to the Women's Institute and similar groups, who am I to turn down such high-quality baking!

For years I have known with all my heart that I am right – that my view of the world is correct and that those who kill animals for pleasure, or have others do the killing for their pleasure, are wrong. My social media feeds confirm that I am right, for the most part. And prove to me that I am, in fact, only half-hearted in my dedication to save animals and be good.

And so the personal bubble grows in strength, those within it hearing only amplified versions of what they already think. In this book I am going to take the very challenging step outside this bubble, and will poke my nose into the surrounding bubbles. Why? Because I have found arguments for killing animals that I cannot 'win' from within my bubble. And as I write that I am reminded of the famous statement by Joseph Joubert, the 19th-century French moralist: 'The aim of argument or discussion should not be victory, but progress.'

The essence of science is not that it knows everything, but that it continually challenges everything. It is not just what we think, but how we think that needs challenging.

What to do about principles? I was brought up to believe that I should have principles and that I should stick to them. But are our principles informed by prejudice? Or maybe it is just the case that I have principles but you have prejudice!

Clearly there are no hard-and-fast rules on what constitutes the correct treatment of non-human animals; there are dramatically different views, driven by culture, religion and money.

Since the dawn of philosophical discourse we have considered our relationship with the non-human world. This might seem like a very esoteric place to begin to look at the very practical issues being discussed here, but I think it is important to at least have a grounding in the thinking that has led us to where we are now.

For me, an essential guide has been Lori Gruen's 2021 book, *Ethics and Animals*. She says that 'it is against the animal that we define humanity'. She contrasts this to earlier thinkers, such as Aristotle, who placed animals on a lower level than humans because of their lack of reason. In fact pretty much all the early writers and thinkers from the western traditions, with the exception of Francis of Assisi, saw animals as a means to an end.

Leaping forward from Aristotle to the seventeenth century, it was René Descartes who most famously espoused the difference – humans have souls, while animals do not and are essentially automata. All of this is important because it brings us to the issue of who or what has 'moral standing'. If you can split life so easily into beings with and without souls then it is straightforward to deny moral standing to non-human animals.

Moral standing is essentially a sophisticated way of deciding whether someone or something 'counts' – do we need to consider the way we treat other people? Yes, I think most of us would agree with that. But how about non-human animals?

I think we should give moral consideration to other species – but does that mean I consider all species to be equal in standing? No. In the same way, I imagine I would tend to leap to save my children before a stranger if they were both in equal danger, and I would rush to save our family dog before one with whom I had no connection. We create hierarchies of care. But these do not remove the fact that I would give moral consideration to the stranger – human or dog.

Gruen helps to give this perspective – she writes, 'Some philosophers have suggested that the wrongness of acts of wanton cruelty does not arise from the *direct* harm the act has caused on the animal victims, but rather that such actions are thought to be wrong because they reflect the type of character that often allows a person to engage in unethical behaviour towards humans.' She singles out Immanuel Kant, who considered that 'irrational animals' did not warrant moral consideration. He argued that, for example, the shooting of a dog that the owner thinks of as worthless does not breach any duty to the dog, as the dog has no moral consideration. However, it does damage the qualities of the owner, who should act in a humane and kindly manner.

I am not a philosopher, and have found that dipping my toes into their world has been both daunting and rewarding – in particular when I have had the good fortune to talk to my friends who come from that discipline. They have stepped in to help clear the fog.

I have known Paula Casal for nearly 30 years, and was always rather intimidated by her obvious intellect. She is now a professor and co-director of the Center for Animal Ethics at Pompeu Fabra University in Barcelona. She has, over the years, been working on many projects, including one with Peter Singer, the philosopher who published the massively influential book, *Animal Liberation*, in 1975. I got a copy in the mid 1980s and while I read the words, in retrospect I don't think I really understood the importance. All I got from it was that clever people had reached a similar conclusion to mine, therefore my conclusion must be right.

On rereading some of the text, with encouragement from Paula, I am much more aware of the subtlety and importance of Singer's work. For example, while it might have been a founding text of the Animal Rights movement, Singer himself rejects the concept of 'rights', instead saying that the interests of animals need to be considered because they can suffer, and that the use of utilitarian principles should lead to a minimisation of suffering.

Utilitarianism sounds rather mechanical – a case of weighing up joy and suffering to create the greatest happiness in the outcome. In the early nineteenth century the philosopher Jeremy Bentham defined the 'utility' that was to be maximised as 'that property in any object, whereby it tends to produce benefit, advantage, pleasure, good, or happiness ... [or] to prevent the happening of mischief, pain, evil, or unhappiness to the party whose interest is considered.'

As with all gatherings of clever and wise people, the arguments as to whether this is a sound view to take into the world will continue while they still have breath, but I feel that as a basic premise, minimising suffering is a good starting point.

Singer takes us further, though. And I was fascinated by the idea of speciesism. He describes how it favours humans: 'the racist violates the principle of equality by giving greater weight to the interests of members of his own race, when there is a clash between their interests and the interests of those of another race. Similarly the speciesist allows the interests of his own species to override the greater interests of members of other species. The pattern is the same in each case.'

This argument is at the heart of the trouble with human exceptionalism; the idea that we as a species are special, that taxonomy can be simply broken down into 'us and them' – where 'them' is everything else. But there are clear problems with the simplistic notion of human superiority. It is most usually based on capacity – that because we can do things that other species cannot, we have a cognitive experience superior to other species too.

If you chart the debate over our human, exceptional qualities, you will find goalposts being shifted with indecent haste. The defining notion of humanity was once that of tool use, so much so that there was a call to name our species *Homo faber* – man the maker. But Jane Goodall, among others, put a spanner in the works with her observation of wild animals doing just that – using tools. Her work showing that chimpanzees were able to fashion tools to extract termites ended that debate. So the goalposts were shifted – humans were deemed to be the only animals that use 'tool kits', where tools can be used for multiple functions.

But again, those out in the field, spending time with wildlife and watching and learning, saw that this too was not enough. Chimpanzees were seen using stones to crack nuts and then twigs to extract the goodness within. Other chimpanzees were seen making leaf sponges to collect water, and then, when the water was too low, using sticks to extend the reach of the sponge. It is not just mammals; New Caledonian crows are also adept at tool use and have been seen using a short stick to retrieve a long stick, which was required to extract a food reward.

So the goalposts were moved again. Perhaps, suggested those keen to establish human exceptionalism, it is the preparation and collection of tools before they are needed that is unique. Again, no. Both chimpanzees and crows have been witnessed working in this manner.

How about a different gauge? Language is another totem – and there is a degree of uniqueness among the primates in the ability of humans to physically make the noises required for speech. But then, there are many humans who cannot speak as others, and for many of them alternatives have been developed, such as sign language. The results from teaching common chimpanzees, bonobos and gorillas sign language are amazing. And more than learning words, the chimp Washoe, originally captured for use in the US space programs, was taught American Sign Language in the late 1960s. On seeing a swan for the first time she signed 'water' and 'bird' – displaying a further level of intelligence than simply learning by rote. Harvard psychologist Roger Brown described this as 'like getting an SOS from outer space'.

Koko the gorilla was also able to create new words through combination, and the orangutan, Chantek, was able to learn 150 signs. Even within this world of research the bar kept on being raised as those clinging on to human exceptionalism fought back. The question of whether these apes learned to use what we would define as a language annoyed the purists. Linguist Noam Chomsky argued that it was irrational to try and teach linguistic skills to animals. He described it as 'like trying to teach people to flap their arms and fly.' Though I think he might just have been smarting from finding out that one of the clever subjects had been named Nim Chimpsky.

Another way of separating us out was to look at whether non-human animals have a 'theory of mind' – that they recognise themselves as an individual, and recognise other individuals as having emotions, perceptions and thoughts. A further question that is raised in the effort to identify our exceptional nature is whether non-human animals can behave ethically. Well, there is evidence, though contested, for the theory of mind – the contestation comes mostly from bar-raising. And as for behaviour, ethologist Marc Bekoff writes, 'Animals not only have a sense of justice, but also a sense of empathy, forgiveness, trust, reciprocity, and much more as well.'

My motivation for pursuing this line of thought is to deal with the polarisation – to demonstrate that it is impossible to define absolute exceptionalism. All it takes is one example to undermine that cause. And when that has happened, acceptance of the moral consideration of non-human animals comes a step closer.

Angeliki Kerasidou is another Oxford philosopher friend – partner to the ecologist and writer Tom Moorhouse. I remember the first time I realised quite how daunting her thinking was when I had invited them both around to my house to drink wine and eat cheese with George Monbiot – the *Guardian* columnist campaigning journalist. She sat quietly, listening, while we men nattered away thinking we were being wise, until Angeliki spoke up. I can't even remember the details – just that we were all rather reminded of our place.

We met recently for a walk around my local park. She was pushing her pram and I felt I would have a chance – she had just had a baby; I could surely keep up with her thoughts! It was fascinating to be introduced to the basics – it really was Philosophy 101. Was I prejudiced when it came to the way people treat animals? That was the starting point – were we all pre-judging everything, people like me who felt revolted by killing, and those for whom it was a delight? Proving that the baby had not dulled her wits, Angeliki immediately criticised my conceit that I was exhibiting prejudice.

'This is moral intuition – the emotion you feel when thinking about this is a combination of your cognitive appraisal – what you think about consciously, and the simultaneous interaction with somatic perceptions; how your body is reacting to the situation. Look, there are some things which seem universal – like killing is bad. But even that can be considered through different lenses – so are you going to consider whether something is good, or bad? Or are you going to follow rules because something is right, or wrong? This is where the book *Would You Kill the Fat Man?* will be your friend.'

This eccentrically titled book is the work of David Edmonds and it revolves around Trolleyology – which is slightly more understandable when you consider that the American trolley is what we call a tram. There are many subtle thought experiments, but the ones that stuck with me are these.

Five people are tied to a tram track – there is a runaway tram on the way that will kill them. You are standing by a lever that you can pull and direct the tram onto a spur, saving the five, but there is a single person tied to that track, who will die. Do you step away, do nothing and let five people die? Or do you act, and in so doing, ensure the death of one person?

The fat man features in a different iteration of this problem. Now, there are still five people tied down, unable to escape, and you are observing the runaway tram from a bridge, while standing next to a large man. And you know that if you push him, his bulk will be sufficient to block the tram, saving the five, but killing him. What do you do?

What Angeliki was introducing me to was the idea of good (killing one in order to save five) being better than right – that the consequences are more important, and this is at the heart of utilitarian thinking. The alternative, that right trumps good, that rules must be obeyed (so letting the five die because it is wrong to kill), is known as deontological, or normative, thinking.

I think I am a utilitarian. But a deeply troubled utilitarian. Killing for conservation troubles me, because the action of killing animals is deeply upsetting. But the consequences of that killing, should they be done for good reasons, will result in a threatened species being able to continue to live. Now, whether the conceit of a species having specific value requires scrutiny, we will come on to later.

Actually this is all a fancy way of saying that the basis of this book is the collision between heart and head. And please trust me – at no point has this been easy. There may well be conclusions I draw that some people will find upsetting. I assure you that reaching those conclusions will have been upsetting for me as well.

Though the research has not been without entertainment … there is a subspecies of philosopher who largely never leave the chaise longue on which they await wisdom – and this leads to what common folk like me tend to refer to as an outpouring of utter bollocks. At this point I should indicate that the idea that follows is not a flight of fancy – despite occasionally taking to the stage with a very niche act of hedgehog-based stand-up comedy.

At first the philosophising really did seem like a joke. The reasoning goes thus; we recognise that animals are sentient, that they suffer pain and fear, and we, aware of this, should do our best to remove pain and fear from their lives. So far so good. The next idea is also true – that in nature, wildlife suffers pain and fear through the process of predation.

And what leap of the lounging mind do you think happened next?

Well it is not a leap that is exactly new. Back in the eighth century BCE, the prophet Isaiah wrote some of the most important words that appear in the Bible. He called for the abandonment of war. 'They shall beat their swords into ploughshares, and their spears into pruning hooks: nation shall not lift up sword against nation.' This statement spawned some of the most vital acts of direct action, centuries later in 1980, thanks to the Ploughshares Movement – I heartily recommend looking them up. But Isaiah then went a step further, by espousing universal veganism, 'The wolf also shall dwell with the lamb, and the leopard shall lie down with the kid; and the calf and the young lion and the fatling together; and the little child shall lead them. And the cow and the bear shall feed; their young ones shall lie down together; and the lion shall eat straw like the ox.' (Isaiah 2:4 and 11:6–7)

Anyone with an ounce of common sense will see a few obstacles to this – not least with carnivores' dentition and gut arrangements. But this has not stopped modern prophets re-animating this idea, including some at the University of Oxford. Jeff McMahan has been a professor of moral philosophy at Corpus Christi College since 2014 and has written extensively about many areas of suffering with a deep wisdom.

Yet in conclusion to a 2010 essay in the *New York Times* on this subject he wrote,

> *It would be good to prevent the vast suffering and countless violent deaths caused by predation. There is therefore one reason to think that it would be instrumentally good if predatory animal species were to become extinct and be replaced by new herbivorous species, provided that this could occur without ecological upheaval involving more harm than would be prevented by the end of predation … I am therefore inclined to embrace the heretical conclusion that we have*

> *reason to desire the extinction of all carnivorous species, and I await the usual fate of heretics when this article is opened to comment.*

Now his measured observations did not go unnoticed and there is a rash of young thinkers who have jumped onto this toboggan of logic and headed towards the crevasse of doom … Oscar Horta wrote that 'most sentient animals who come into existence in nature die shortly thereafter'. He has argued that animals suffer due to a number of natural causes (including starvation, disease, conflicts, and others) and that since they are sentient beings, this gives us reasons to help them.

Horta's paper is so filled with straw men that it is a fire hazard. His conclusions start reasonably enough – that non-human animals should be morally considered. Moving from there the assumption is made that most people are unaware that life in the wild is tough, which I would question. Then he argues that any opposition to intervention is foolish as we already intervene, wildlife rehabilitators being a case in point. From here it is a simple step to conclude that we need to modify species and the environment in the journey to a herbivorous utopia. Possibly hinting that this is an extensive joke, he concludes that it is important 'that this issue becomes a respected one that is taken seriously in academia'.

So maybe in this spirit I will be able to inject some good humour into the next few hundred pages. As George Bernard Shaw said, 'If you're going to tell people the truth, you better make them laugh, otherwise they'll kill you.'

Our history as a species doesn't give us much hope for our ongoing relationship with the natural world. I have friends who praise the flawless relationship Indigenous people have with the wildlife and the land they share, who wax lyrical about the time before now when we were so much more in touch with nature. These people are often very angry at the way modern life has caused such damage to the world. It is easy to be sucked into such a simplistic world view and believe that surely things were better back then ... though personally I am rather fond of modern surgery, antibiotics (used sensibly, of course), sanitation and my iPhone.

At the New Networks for Nature gathering in Bath in 2021, one of the panellists made such a comment about the halcyon time before, well, take your pick. His thrust was that science had driven a wedge between us and a way of living with nature that was more natural and in all ways better for us and it. This got me twitchy in my seat and at the very moment the proceedings were opened up for questions I was like Hermione Granger at her most insistent.

I had the advantage of being in the front row, so I got my chance to politely ask how he could square the circle of our history with his fantasy. Because, as a species, we have one really consistent feature. We cause extinctions. We are killers and ecosystem engineers and we have a fundamental flaw in our make-up. *Homo sapiens* evolved 300,000 years ago at a time when planetary limitations were unknown, when surplus was a delight to be exploited. Now we understand these things – well, some of us do – that perpetual growth on a planet with finite resources is clearly impossible and that our actions do indeed cause extinctions. But while our brains can understand that, we are, like a fox in the chicken coop, acting on instinct, in our case driving the liveable planet towards destruction. At times I think we should rename our species *Homo occisor* – man the killer.

Wherever humans have gone since some of us departed our ancestral home in Africa 90,000 years ago, we have left destruction in our wake. Look to the evidence from the end of the Pleistocene in North America. Between 10,000 and 12,000 years ago, as the ice sheets retreated, modern humans crossed Beringia, the land bridge from Siberia to Alaska. Now, these were not the first people in the Americas. There is evidence from the Chiquihuite Cave in Mexico that suggests human tool use from up to 33,000 years ago. But it is thought that humans were fairly thin on the ground in the Americas until an additional arrival at the start of the Holocene.

As humans progressed south, so a series of dramatic extinctions of megafauna took place. Beavers the size of bears, sloths the size of elephants – gone. Yes, there is some debate as to what impact the changing climate had, but evidence of skilled hunters of the time, who left behind distinctive projectiles, has been widely excavated throughout North America. Smaller mammals and reptiles were not impacted in the same way, suggesting that selective hunting for (probably naïve) megafauna was the cause of these mass extinctions. This also happened to the woolly mammoths and cave bears, among others, of Europe, though over a longer time.

So the romantic vision of early humans living in harmony with nature is a myth. As now, we over-hunted prey and zealously protected what remains from the competition.

It is interesting that the megafauna of Africa, where humans evolved, still exists. It was the speed of the arrival of humanity across the rest of the world that was the shock to the naïve wildlife of Eurasia and the Americas; these amazing creatures did not have a chance against the precocious apes.

We have not stopped. So much so that we are being written into the epochs of the planet – the Holocene, the kind and gentle epoch that has allowed the rapid expansion of humanity, now gives way to the Anthropocene in which our mark is being indelibly left. Archaeologists in a million years, should there be any left, may give it a different name, but they will find a narrow stratum of plastic, chicken bones and strange radionuclides and wonder what on earth happened then.

The most obvious impacts for which we have more historical evidence comes from the arrival of people onto previously uninhabited islands. And of these the most famous story is probably that of the dodo. But it is not just over-hunting from people that drove this large flightless, and apparently not very tasty, bird to extinction – it was also our fellow travellers, the pigs, cats and monkeys previously not resident on Mauritius, that finished them off.

Some of these extirpations have entered folklore – at least with the sort of folk I know. Though it seems that the story of Lyall's wren – a flightless bird found on an island off New Zealand – being exterminated by the lighthouse keeper's pet cat, Tibbles, within a couple of weeks of it being identified, might not quite be as true as its quality warrants. Cats had already killed off the rest of the species on the mainland, and there were many feral cats on the island to deliver the coup de grace.

How do we undo the damage? When we humans have unleashed a new species on an unsuspecting fauna, when we have transformed ecosystems so that previous balances become undone, how do we fix the problem? Should we fix the problem? Should we play god, or should nature be left to take its own course?

I think it is important that we take a moment with every one of these decisions to pause, and to remember we are setting ourselves up as judge, jury and executioner – or at least supporters of those that are. We are fully embracing the idea of human exceptionalism – the idea that we, as a species, are able to play god with the rest of the world.

But playing god is never going to be straightforward. The sparrows of China were declared an enemy of the state in 1958 by Chairman Mao. He had calculated that by killing a million sparrows, grain to feed 60,000 people would be saved. So a nationwide eradication attempt was launched; 2.8 million sparrows were killed in Shantung Province alone.

The result? A catastrophic failure of the was not just rice harvest. Because what Mao had failed to recognise was the staggering number of insects these usually granivorous birds feed their young – plant-eating insects that were left to ride unhindered through the nation's larder. The campaign was quietly dropped and sparrow numbers were back up to where they had been within just three years.

Perhaps this example is all that is needed to persuade those in authority of the importance of having an ecologist oversee everything. But even with ecologists aplenty we are left with the problems that challenge the head and the heart as they fall out with each other. And we are confronted with the rather unedifying disappearance of reasoned debate. I do not mean that there was ever a golden era when reason ruled – even the Ancient Greeks relied on systems of slavery to enable the elite to spend their lives in debate.

There has been a fictional democratisation of debate through the rise of social media – one where the opinions of a few have received more airtime than they possibly deserve, where algorithms promote dissent by encouraging, no, in fact demanding, outrage. It has got to the point that UK government minister Michael Gove said, 'I think the people in this country have had enough of experts...'

I undertook an accidental experiment on Twitter (now known as X) at the start of 2023. At roughly the same time I posted a short video of the robin in my garden taking a mealworm from my hand, in slow motion, and also a tweet in which I shared my unhappiness that a company was purporting to sell hedgehog poison – among many other species-specific poisons. That is in itself a long story – but which tweet was viewed well over a million times? Certainly not the cute robin. Outrage is the motor that drives the dopamine hit of acceptance online. For some, the greater the reach of a social media post the greater the financial returns – a level up from dopamine. Which again leads to more outrage and inflammatory comment, in turn further emptying the moderate and considered space in the middle.

I hope that what follows, as I meet with ecologists at the cutting edge of some of the most pressing conservation challenges, will avoid too much outrage, and encourage us to pause and think and realise how any cull of the wild is very complicated.

CHAPTER 1

Hedgehogs

Let's start with hedgehogs.

Given my history it is rather difficult to avoid hedgehogs. Having spent more than 35 years working with them – studying them, campaigning to protect them, writing about them, talking about the love I feel for them, doing stand-up comedy about them and even getting a (mid-life crisis) tattoo of one – it is logical to assume that I would vigorously oppose any deliberate attempts to kill them. Yet in 2010 I wrote an article supporting the cull of hedgehogs.

Before you come at me with pitchforks and flaming torches, please, let me explain.

As I have already said, ecology is not simple. One of the principles by which science survives is that we accept it is just the best way we currently have of explaining what is going on around us. Everything needs to be tested because we cannot rely on 'belief'. Things that I thought were true have turned out not to be true. And in all honesty, when presented with new evidence we have to reconsider our thoughts. Science is always changing. The comedian Dara Ó Briain summed this up perfectly: 'Science knows it doesn't know everything, otherwise it would stop.'

And this is where my relationship with the hedgehog takes a strange and, to many people, unexpected turn. The acclimatisation societies of New Zealand were founded in the 1860s by European colonists as they attempted to soften the psychological disconnect of an unfamiliar world. What better way to make it feel more like 'home' than by requesting a collection of familiar animals?

For example, the Otago Acclimatisation Society was in receipt of £500 a year from the state government to import familiar British species, mostly small birds. This is why starlings, blackbirds, house sparrows and chaffinches are common around Dunedin today.

Not all introductions were so inoffensive. Pheasant and quail numbers increased to the point that poisoned grain was used in their control; when that did not work they moved onto releasing stoats, weasels and ferrets. I know that hindsight is a glorious lens, but even so, could the potential for this to go wrong not have been foreseen?

As for the hedgehogs, the *Canterbury Press* announced their arrival in New Zealand on 4th October 1869.

> *A pair of hedgehogs, the first imported into the province, have been presented to the Acclimatisation Society by Mr D. Robb, the purser of the* Hydaspes. *As the food of these animals consists principally of insects, they will no doubt prove of invaluable service in keeping down the grub. Anything which can help to destroy the rapidly increasing farmer's pest, will be a great boon to the province. Whether the amount of good done by hedgehogs is counterbalanced by their partiality for eggs and young birds, is a point upon which our Acclimatisation Society has as yet arrived at no definite conclusion.*

Fascinating to see already the note of concern – however, it seems that this was directed more towards chickens and introduced game birds, rather than any native species.

Over the following years there were more releases as the attempt to help the European settlers settle continued. There are records of introductions in 1871, 1885, 1890 and 1894. By 1972 hedgehogs had reached their maximum coverage, occupying most of the available habitat on both North and South Islands.

Pat Morris, my friend, mentor and the first academic to take the study of hedgehogs seriously in the United Kingdom, undertook a roadkill survey in New Zealand in the summer of 1987. This seemingly gory pastime is actually a really important way of getting an indication of the state of the hedgehog population. Travelling with his wife, Mary, they recorded 636 dead hedgehogs during their mammoth

9,966km drive. Having looked at a map of their travels, however, I am less inclined to feel sympathy for the long drive, and more envy at their coverage of this magnificent landscape.

The revelation that the number of hedgehogs found dead on the roads was similar to that found in a survey Pat did in the United Kingdom, where there were 10 times the number of vehicles, gave a good indication that this survey technique is a measure of hedgehog numbers in the environment, and not a measure of cars. Obviously this only works on certain roads at certain times – as we found in the United Kingdom during the 2020 lockdown when the number of roadkill hedgehogs was dramatically reduced.

For many years, the ex-pat hedgehogs were considered to be the most benign of the incomers. They were a popular addition to gardens of suburbia. They did not come under the watchful eye of the conservationists who were trying to undo the damage caused by rats, stoats and possums, to name but a few. In fact the most notable occurrence of the hedgehog was within the delightful comic strip produced by Burton Silver for the *New Zealand Listener* – featuring the philosophising woodsman, Bogor, and his friendly marijuana-munching hedgehog sidekick!

However, there were some concerns, raised back when the hogs were becoming established. For example, on 26th February 1915, in the *Hawera and Normanby Star*, there was this note::

> *A plague of hedgehogs is evidently as bad for the poultrykeeper as an incursion of weasels. Most farmers know that hedgehogs have a profound partiality for eggs, but until recently they considered that their well-grown chickens were safe from them. But Leeston is just at present suffering from the inroads of an unusual horde of hedgehogs, and the lives of the domestic fowl are precarious (says the* Christchurch Sun*). Several farmers have had their flocks of*

> *egg-producers seriously depleted, and one lost 10 young turkeys in a single night. There is no doubt that hedgehogs are the marauders, for several which have been killed have had traces of feathers round their mouths.*

I do wonder whether these hedgehogs might have been scavenging. While it is entirely plausible that one might kill a chicken, if the bird was not able to get off the ground, they are not noted for their killer drive. The reason that foxes, for example, are so often reviled, is the fact that they will seem to kill more than they can eat. This is because the fox will kill what it can, which in the wild will be the sick or injured prey. The infantilising process of domestication reduces the capacity of poultry to escape, as does the housing. This in turn leaves the fox in a quandary – why are the meals not flying away? They must be weak, so I will keep killing, considers the fox, who, if left to its own devices will carry away and stash the bodies for later consumption. But they are all too often interrupted by an understandably irate farmer.

But hedgehogs – they just don't do that. They will mooch and munch – following their sensitive snout from food to food. On 26th February 1934, the *Southland Times* was back on the same topic.

> *The menace of the thousands of hedgehogs which he said were in the district was referred to by Mr W. H. Moyes at the council meeting of the North Taranaki Acclimatisation Society last week. 'Hedgehogs are going to knock us,' Mr Moyes said. He went on to relate how he had disposed of some hedgehogs which he had encountered. Three which he had imprisoned one evening by placing a box over them had burrowed their way out by the next morning, when he had intended to dispatch them. They had killed about 22 pheasants. A member referred to the value of hedgehogs as destroyers of slugs and garden pests.*

I find it fascinating how the attention remains purely on the impact of hedgehogs on other imported species. One can only imagine what the hedgehogs were getting up to in the undergrowth and among the naïve fauna of these beautiful islands.

While I refer to them as beautiful, I have never actually been to the islands of New Zealand, and am in fact rather nervous of the place – I fear that if I go I will never want to leave. Peter Jackson did not help. His imaginings for the *Lord of the Rings* films could have been funded by the New Zealand tourism board – though it is possible that hordes of orcs might put off some less intrepid visitors. All that beauty, and hedgehogs too!

To be fair, I have managed to have time in other gorgeous and beautiful places. Take the Uists, for example, and another unfortunate hedgehog story. These islands, on the western reaches of the Outer Hebrides, have really only St Kilda in their way before you reach Newfoundland. Yet their potential bleakness is rather subverted by the tender caress of the Gulf Stream. In spring the Machair, a globally rare low-lying fertile plain, bursts into life.

The diversity and abundance of flowers is just one indicator of the special character of this land. The Uist Machair is also home to important populations of wading birds. A survey in 1983 found that it held 25 per cent of the United Kingdom breeding population of dunlin and ringed plover. The survey was repeated in 1995 and found that dunlin, ringed plover, snipe and redshank had all declined dramatically.

Ecologist Digger Jackson was recruited to look into this problem. He was concerned that the introduction of hedgehogs in 1974 to the south of South Uist might be the cause for the declining breeding success that had been noted over repeated surveys. The evidence seemed to be strong as the bird populations suffered in the wake of the advancing line of hedgehogs. Shorebird breeding numbers declined by 39 per cent between 1983 and 2000 where hedgehogs were present, but increased by 9 per cent on North Uist, where they were absent.

The RSPB now state that their research found hedgehogs were 'the main predator of wader eggs,[and] to be a key factor in causing breeding success that is low enough to contribute to population declines in wader species that are susceptible to hedgehog predation, such as Dunlin and Redshank'.

I remember hearing Digger talk at a conference at the beginning of the century about this and feeling my heart sink. Clearly, my beloved hedgehogs were in the wrong place at the wrong time and would have to be removed. There was a real risk of these birds being wiped out. I had seen this happening before. My first hedgehog research back in 1986 was to investigate a very similar tale – of hedgehogs brought to the most northerly Orkney island, North Ronaldsay. On this small island, it was the Arctic tern population that was causing particular concern. As hedgehogs increased, so the terns dwindled. In this instance, hedgehogs were one of many challenges the birds faced.

I estimated the population to be around 500 hedgehogs and soon after I left the island, in the late summer of 1986, the North Ronaldsay Bird Observatory organised an airlift. This made the news as islanders were encouraged to scour the small, five-mile-long narrow island. Hedgehogs were boxed up, flown to the mainland from where they were put onto the train and deposited with volunteers all over the United Kingdom who had applied to receive the island refugees.

This was not a robust study, there was no follow-up, but it seemed to do the trick. When I returned in 1991 to repeat the survey, the number of hedgehogs was massively reduced and soon after slipped from view and memory, though they had not completely gone, as their reappearance in later years showed. However, the time they were invisible did not correspond with a resurgence in the breeding success of ground-nesting birds. This suggests that there were additional factors affecting the ability of these birds to thrive, such as changing farming practices and dwindling fish populations.

However, elsewhere hedgehog numbers had continued to grow unchecked. In 2003, Jackson estimated there were 7,000 hedgehogs on the Uists, which presented an enormous problem. To remove most of them would not be enough, as the situation was started by just a handful. And there was a pressing additional concern – the EU Wild Birds Directive required that member states take special measures to conserve the habitats of rare bird species. These areas should be classified as Special Protected Areas (SPAs) – and there are two in the Uists. This meant the Scottish government was obliged, under international law, to ensure that the SPAs were not degraded in any way, and that the birds had to be protected.

To that end an old friend of mine was recruited to design a way out of the mess. Nigel Reeve, a brilliant ecologist with a keen interest in hedgehogs, presented a feasibility study that outlined a way of assessing whether the translocation of Uist hedgehogs back to the mainland (as had been done in Orkney) was both practical and humane. He asked me to supply him with the evidence from my work up in Orkney, which I was happy to do, but I was totally tied up with other work, as well as getting married and having a baby, so I did not keep an eye on what else was happening.

What I was unaware of, and this proved to be rather significant, was that the newly formed Uist Wader Project (UWP) had rejected Nigel's thorough report and contracted James Kirkwood, who was at the time the chief executive of both the University Federation for Animal Welfare and also the Humane Slaughter Association, to interpret the data differently.

Kirkwood's report contained three options, after dismissing catching hedgehogs on the islands and moving them to the mainland as inhumane. These were: to catch hedgehogs and keep them in some sort of zoo on the islands until they died; to utilise as-yet-undiscovered contraceptives; or to kill them. It came as a great surprise when the boss of the Humane Slaughter Association recommended that the hedgehogs be humanely slaughtered.

But as I mentioned, this was all information of which I was unaware. In 2003 news stories began to appear about the plans to kill the hedgehogs on the Uists to protect the birds. These began to register through the fog of first-time fatherhood – I was rather entranced by a small person called Matilda.

The cull, as the RSPB and its consortium of the Scottish government and Scottish Natural Heritage (SNH, now Nature Scotland – or NatureScot), preferred to call it, was due to begin on 6th April 2003. For some reason I had it in mind that it actually began on 1st April, just because of the farce that followed.

Leaving my wife Zoe in charge of Matilda, I headed up to cover the story for BBC Radio 4's *Natural History Programme*. Unaware that his work had been rejected, I went sure that Nigel's feasibility study had uncovered something new – that the obstacles to translocation had been too great. I trusted him as a scientist to have been rigorous and true and if he had found reasons why culling was the best option, then so be it.

I was not alone in thinking this an interesting story – most major news organisations had decided that it was something to cover. SNH's spokesperson created a fascinating and newsworthy scenario. He said that the expert opinion was that if the hedgehogs were to be moved alive, they would suffer 'slow and lingering deaths' when released onto the mainland.

He also told us of the gang of radical animal rights activists on the island who were trying to obstruct the work of the conservationists – and because they were present we, the press, could not see the faces or learn the names of the people who would be finding the hedgehogs and taking them to a secret location for their lethal injection, for fear of making them a target for retribution.

The trouble was that I had already met these dangerous animal rights activists; they were in fact a loosely affiliated bunch of animal lovers under the banner of Uist Hedgehog Rescue and were as gentle and non-threatening as can be imagined. Their aim was to rescue as many hedgehogs as possible. They had support on the islands; the collected hedgehogs were to be cared for until there was a van-full, and then they were to be taken to the mainland on the ferry to be released after a full health check from Hessilhead Wildlife Hospital.

Better still, these 'radicals' already knew where the killing shed was, and posed for a photoshoot with a banner outside it. And they had been chatting to those doing the cull, making sure that they worked in different parts of the islands to avoid a clash with two teams diving for a hedgehog, whoever got there first determining its fate.

The day after the cull had started, I went back to the rescuers and shared what I had heard about the dangerous animal rights activists on the island. Luckily they found this amusing. And I also had a closer look through the history of the discussions about what to do with the hedgehogs. It was fascinating. Not least was the revelation of what had happened with Nigel's report and the advice from Kirkwood – which was used to justify the cull.

The claim that hedgehogs would suffer slow and lingering deaths was, in part, based on some of my own research. I was furious – the study I had done was looking at the release of naïve hedgehogs into the wild, having been taken into care as infants and having spent the entire winter awake in a wildlife hospital. The high mortality – five of the 12 died – was used as an argument against translocating healthy adults to the mainland. But not one of my five died a slow and lingering death. All five suffered fast and violent deaths. Two were run over and three were eaten by badgers.

My role shifted over the next few months from reporting to supporting. I became a part of the Uist Hedgehog Rescue team, helping raise money and awareness for the rescuing volunteers. We instigated a bounty scheme, encouraging the islanders to get engaged – hedgehogs delivered to rescuers were met with a small, and then increasing, payment. We got celebrities involved; Queen guitarist Brian May helped fund the work. Sir Tim Rice was among many with land who offered to accept rescued hedgehogs. Even the politician Ann Widdicombe became a vocal supporter.

The more I learned about the path towards the decision to kill, rather than move, the hedgehogs, the more my hackles (prickles?) rose. The rescuers were dismissed as either a bunch of bunny huggers who were too soft to appreciate objective science, or as dangerous radicals, but the reality was very different. Some of the top mammal scientists in the country were working together, and were being ignored.

In 2004 I was invited to the European Hedgehog Research Group meeting in Münster, Germany, to talk about the hedgehog cull and our attempts to get it stopped. The meeting was, as they always are, about so much more – and it was a fascinating chance to meet researchers from all over Europe and learn about the work they were doing.

But perhaps the most significant moment came when we listened to Chris Jones from Landcare Research, the government-owned research institute from New Zealand. He had come over to talk about the situation they were facing, about their need to kill hedgehogs. I had just been through this – of *course* there was no need to kill hedgehogs, and anyway, what I had read about the New Zealand situation suggested that hedgehogs were a benign alien.

And here is the irritating thing about the scientific approach – if you are willing to be an honest listener/observer you might end up having to change your mind. It is the exact opposite of 'faith' – I am always amazed that when presented with a clear refutation the faithful will often harden their opinion.

Listening to Chris, it became clear that there was a serious problem in New Zealand and that something had to be done. But still, the option of killing – it just seemed wrong. What had the hedgehogs done to deserve that sort of treatment? They did not ask to be there. They were just behaving as hedgehogs behave; it is our mess, our fault.

Which is great, if you are still in primary school. But as you grow up you have to take responsibility for the mess that is made, even if you are not immediately responsible. That is the shorthand version of what was said.

I caught up with Chris late in 2022. He has moved on from his first love of fieldwork and is now managing many other projects – 148 at the last count. Landcare Research has also moved with the times, and is now known as Manaaki Whenua, in acknowledgement of the growing recognition of the importance of Māori culture when looking to manage the land. It is far from an affectation too – talking with Chris on Zoom it was clear that the more he learned of Māori, the more he realised the congruence of his instincts and much of their culture.

The biggest of the projects that Chris is involved with running is the nationwide Predator Free 2050 – an ambitious piece of work aiming to remove rats, mustelids and possums from the whole country. For now, though, I wanted to concentrate on what had been going on with hedgehogs since we last spoke, nearly 20 years ago.

'One of the most fascinating things about the hedgehog is how deeply ingrained they are with certain members of the public,' he said. 'There is this very Euro-centric perception of the hedgehog – in particular among older people with closer ties to Britain. They still have a strong feeling for the animal. But there has been a growing awareness of the biodiversity impact over the last fifteen years or so.'

An indication of how things have changed comes from an article written by the writer, scientist and hedgehog advocate, Rob Brockie. In 1999 he wrote a piece in the *New Zealand Geographic* magazine, beginning with this amazing, and quite relatable, anecdote of the perils of fieldwork.

It was 3am and raining heavily. I was wearing a balaclava and gloves and carrying a powerful spotlight. Suddenly five police officers rose to their feet from their hideout in tall, sodden grass under the pines. They had been following my movements around a Lower Hutt golf course for over an hour after receiving a tip-off from a vigilant citizen who suspected that anti-apartheid activists were digging up the greens where a South African team was scheduled to play the next day. 'Excuse me, sir. Would you mind turning out your bag?' asked one of the officers. Out came a bottle of chloroform and another of alcohol; scalpels, tweezers, pliers, a notebook full of numbers and a motorbike battery. Eyebrows went up. A radio call to HQ. Then, in a tone of exasperation: 'Oh, no! Not again!'

But it is the conclusion that is probably the most important and telling portion of the article. He points out that

… they are guilty of eating the eggs of some ground-nesting birds on coasts or on wide riverbeds, predating native snails and insects, and competing with native animals for food. How great a threat they represent is still unclear.

> *Despite their transgressions, however, hedgehogs enjoy a special place in most people's affections. Small, bumbling, vulnerable creatures, they are, above all, a visceral reminder to be careful while crossing the road.*

As Brockie's article shows, hedgehogs used not to be targeted – in fact those tasked with removing predators from the land would get annoyed with hedgehog by-catch clogging up their traps. They had more important targets to consider. But over time the hedgehog has become the most commonly trapped small mammal predator.

'Look, you are not going to like this,' Chris said. 'But when the Tasman Valley pest control project, which started in 2005, published its records in 2013, well – this is what they showed. The traps that were set along both sides of the valley killed: 15 Norway rats, 139 weasels, 481 possums, 652 ferrets, 1,559 cats, 2,594 stoats and ...' he paused for dramatic effect '...5,813 hedgehogs.'

I had to interject that it was not a case of me liking it or not. I just needed to understand why, and also to know if this was truly the only option.

The Tasman riverbed is near Aoraki/Mt Cook National Park. As I have admitted, I have never visited New Zealand, but Chris certainly paints an enticing picture. This region is known as the Southern Alps, there are blue lakes, there is the country's longest glacier (though this is now rapidly shrinking thanks to other consequences of humanity's action). And it is beautiful, filled with a great diversity and abundance of wildlife including many species found nowhere else on the planet.

Further south in Otago there are vast areas of tussocky grassland. Here it gets really quite cool at night and the lizards, well, they don't move. But they are fat-filled and easy to find for a foraging hedgehog – it seems that pregnant females in particular go for the easy, 'slow food' option.

'Another study showed how just one male hedgehog had 283 weta legs – which equates to around 47 of these flightless crickets ending up in its stomach.' Chris then made the point that was obvious but worth reinforcing – this was just a single hedgehog. 'The environment is full of hedgehogs, munching up native fauna with glee.'

Chris took the rare opportunity to do some fieldwork, opting to research the impact of hedgehogs, on McCann's skink. These lizards are not only endemic to New Zealand, meaning they are found nowhere else on earth, but are also really cute. I know that such a superficial measure should be meaningless, but in the world of wildlife conflicts it does play a part. We can be as objective as we like – but we still have biases at heart. Up to around 7cm long, the little lizards are speckled and striped, and have an important role in Māori culture, being associated with death and the underworld.

Chris showed a clear relationship between increased hedgehog density and reduced numbers of juvenile skinks. It is not just hedgehogs that are a threat; the remains of 49 skinks were found in the intestine of one feral cat. And stoats are known to eat them too. An ecosystem that has evolved without terrestrial mammals is so vulnerable to their arrival.

What were the options? The hedgehogs could be the cause of the extinction of various species found nowhere else on earth. So they had to be separated from them. You could catch all the hedgehogs in New Zealand and put them in a 'zoo' – but I had already dismissed that nonsense from the Uists. There was no time to hope that a contraceptive might be developed, and even if it was, how on earth would it be delivered to the hedgehogs out there in the bush in a manner that did not impact on other wildlife too?

The important discussion taking place was whether there should be an attempt to remove all the hedgehogs, or to try to stop the hedgehogs causing damage to the native ecosystem. The latter would not be without lethal action; it was about creating a *cordon sanitaire* around the most sensitive parts of the islands. This is hardly without precedent. In many instances, there are times and places where the complete removal of an alien invasive species is impossible, but the reduction or prevention of harm by largely eliminating them from the most sensitive areas is achievable.

In New Zealand they use both strategies: to have the aim of removing the hedgehogs completely from specific sites, but to focus primarily on reducing the harm they caused. So this meant identifying areas with high numbers of hedgehogs coming into contact with vulnerable natives. This kept a lot of the action out of the suburbs, where hedgehogs were weaving their charm to good effect, and into the bush.

Techniques for killing hedgehogs tend to centre on traps. The New Zealand Department of Conservation had developed a snap-trap that was working well on many invasives like the stoats, but it was found that the spines of the hedgehogs were acting like shock absorbers, meaning that the kill was not quick and clean. To reduce suffering they developed a new model, the DOC 200, and later the DOC 250, with a more powerful springs. There are warnings to attend training courses before use.

As effective as these traps were, they still required someone to do the rounds, remove the bodies and reset. There is nothing like a good problem to bring out the best in innovation – so welcome the 'Goodnature A24'. This trap used different scent lures to bring stoats, rats and hedgehogs to a chamber. And as they sniffed into it they triggered a compressed-air piston, which delivered the fatal blow. Ideally the deceased would conveniently fall to one side, enabling the automatically reset device to carry on doing its job.

The traps became a bit of a target for animal lovers in the United Kingdom in 2019 when a company started to offer them for sale, particularly to control grey squirrels. Now, if set correctly for squirrels there is no way that they will catch hedgehogs, and make sure hedgehogs remain safe it is possible to buy a guard that will prevent our spiny friends getting into trouble. However, they were sold without this device and the uproar was interesting. The company selling them pointed out that if the mechanical trap was not being used, most foresters would rely on poisoned bait, which would inevitably percolate into the wider ecosystem and cause potential harm to many more species. So was it not better to risk a rare hedgehog fatality to prevent this toxic spill?

This is the sort of 'equation' that features in every wildlife conflict. Another equation that tends to be skirted around features the welfare of the target animals. Because, perhaps counterintuitively, by taking more care, more animals could end up being killed. For example, the Uist cullers tried to make sure that they did not upset the general public by including restrictions on what cullers could do and when. In particular they were scrupulous at avoiding taking females who were lactating. The welfare consideration of not killing females with young was, at first sight, a sensible approach. But there was an unintended consequence. By adding restrictions to the cull it inevitably increased the amount of time required to remove all of the hedgehogs. Which in turn meant that many more hedgehogs would have to be killed.

At the very outset some of those opposing the cull actually argued that, if a cull was going to go ahead then it should be as hard and fast as possible to reduce the number that would be killed. Although hedgehogs are being killed on the island, those remaining will of course continue to breed, and the longer the cull takes, the more will appear, and the more will be killed in total.

How do we solve this equation? Is it better that a few suffer so that more don't die?

In New Zealand, trapping has been the dominant strategy, and it is from this that we have data to see what is happening. It is harder to record the results from poisoning, and poison is very much still used.

'We get a lot of people complaining about the poison we use,' Chris explained in 2022. 'We use 1080 – it is a brilliant poison as it targets mammals. And New Zealand has no native mammals, so there is no risk to the native fauna. Add to that the fact that it degrades quickly in the wild into harmless components, and I can't see the problem.'

Now, I very much doubt that the wildlife managers of the Outer Hebrides are going to be drawn into using poison. But learning about what has happened in New Zealand did not dent my opposition to the killing of the hedgehogs over here. Admittedly my presumptions were tested; it made me more determined to approach the subject in as clear-headed way as possible. And a realisation was dawning – that the insult of being 'subjective' that was, and still is, regularly thrown at people concerned about a course of action is worth digging into.

It is quite possible to do the science objectively. That is what all honest scientists strive to achieve. But in making the decisions around the science, they cannot necessarily bask in that objectivity and claim it for themselves. The decisions as to what to study and the applications of the results are very much in the world of subjectivity. Put simply, deciding that, say, wading birds are more important than the lives of hedgehogs, is subjective.

The reason ecology is so much more complex than the likes of astrophysics is because it involves people. There is no escaping the fact that we are part of the equation. So with the Uists, if no one cared about hedgehogs, if their extermination did not even register a disgruntled tweet, then the return of the islands to their prelapsarian state would be straightforward. But people do care, deeply and passionately, and have, therefore, to be included in the discussion.

For four years two teams of people spent their summers on the Uists looking for hedgehogs. One team killed the hedgehogs they found, the other team took them to the rescue centre before their journey to the mainland. But still there was the uncertainty: was there any reason Uist hedgehogs could not survive life on the mainland?

The first attempt to investigate this was led by Professor Stephen Harris at University of Bristol. In 2004 his team collected 20 hedgehogs from the islands and drove them all the way to Buckinghamshire for a quick health check at St Tiggywinkles wildlife hospital before releasing themwithin just six days of collection, onto land outside Bristol. This rapid translocation, in combination with a spell of dry weather, will not have helped their survival and it is not surprising that most of these hedgehogs died.

I was on the receiving end of a blistering phone call from Harris as he demanded I persuade the British Hedgehog Preservation Society, of whom I had just been appointed a trustee, to get him more money for his research. I had heard about his intemperate nature, I knew that he held grudges – for example he told me that he had thrown my application to do research in his department straight into the bin as I had used someone he considered to be a competitor as a reference – but I had not been directly attacked in that form. In retrospect he was quite entertaining – telling me how he had used his assistant to investigate what I had done with my life and that they had found nothing, absolutely nothing, so how dare I question him.

But it did leave a question – the heart of science, questions. Had his research revealed something new? Did it give credence to the claims of Kirkwood and the cullers? And more importantly, if we could find evidence that hedgehogs did not suffer slow and lingering deaths, would the cull be stopped?

Unfortunately the only trial the UWP would consider was going to cost an eye-watering £160,000. However, we still wanted to check whether there was some deeper problem we had missed, so I set up a smaller research project just to investigate whether the released Uist hedgehogs did suffer slow and lingering deaths.

In 2005 I collected 20 Uist hedgehogs from Hessilhead. They were all health checked and had radio transmitters attached. My job was to try and keep track of them all for a month. It was decided that this would be long enough to show whether they were really suffering from slow and lingering deaths, in other words, starving.

The work was hard; hedgehogs have a habit of dispersing radially from a release point. My results showed that the hedgehogs had not picked up immortality on the islands, but those that died did so for quite understandable reasons. In other words, their behaviour and survival was very much like that of normal hedgehogs. However, the study was not robust enough to change the mind of the UWP. Additionally, problems had been created in the relationship between the UWP and the rescuers, above and beyond the extremely robust research the UWP had required.

Words are powerful. At the start of the cull, when those trying to stop it were described as animal rights extremists, it was probably not anticipated how that would impact future conversations. The reality is that when you have one side painted as beyond the pale, as many in the establishment see animal rights activists, it is very hard for the establishment to then sit at a table with them and talk.

My 2005 results were, however, enough to change the mind of the Scottish Society for the Prevention of Cruelty to Animals, with whom no bitter words had been exchanged, despite their tacit support of the cull. When they read my published, peer-reviewed paper, they decided to drop their objection to translocation as a solution, leaving the cullers without this essential support. It was enough to end the killing of hedgehogs on the Uists.

That did not mean that the programme ceased – all sides agreed that there was enough evidence to 'convict' the hedgehogs of having a very negative impact on the breeding success of the wading birds. What we had disagreed on was the sentence.

From that moment on there was, to a large extent, cooperation. Hedgehogs were found by the Uist Wader Project and looked after on the island before being shipped back to the mainland and released to a new life, away from the lure of easy eggs.

Repatriating hedgehogs to their homeland is also an option that has been discussed with regard to New Zealand. As the story about the New Zealand hedgehogs' fate became more widely known, so the more eccentric visionaries started to contact me. Initially it was with the view of bolstering our flagging population. But the best was when the 'representative' of someone very wealthy embarked on a lengthy exchange with me as they tried to explain why they needed to repatriate the hedgehogs for their 'rewilding' project.

Stepping in to 'save' wildlife for the best reasons can end up causing as much death, and even more suffering, than had they been left to the tender mercies of the cullers. For example, while hedgehog numbers in the United Kingdom are declining, bringing in more from New Zealand will do nothing to halt the decline, and in fact we risk 'feeding the sinkhole'. The problems causing the decline persist, so until we deal with loss of habitat, habitat fragmentation, and destruction of the species that hedgehogs feed on, there is no point. It is an action undertaken to make us feel better, not undertaken with an understanding of the basic ecological issues at play. Currently, if we were to fix the problems hedgehogs face, they would return of their own accord.

Then there is the welfare of the hedgehogs spared execution. They would have to travel back from New Zealand. How? How many dying on the journey would be considered reasonable? There is also biosecurity to consider. Hedgehogs were transported out there when this was not a concern. Now we are well aware of the capacity of wildlife to collect potentially catastrophic pathogens.

The UWP continued working at removing hedgehogs from the islands until the end of 2012. During that time the total number of hedgehogs removed, dead or alive, by rescuers and cullers, was 2,566. If the original population estimate for the number on the Uists, around 7,000, was correct, then computer modelling showed that 1,130 hedgehogs needed to be removed every year for 15 years to give any hope of eradication. The numbers removed would suggest that the end result of 10 years' hard work would actually be an increase in the number of hedgehogs on the archipelago.

To rub salt into this wound, the amount spent by the UWP from April 2003 to January 2011 was £1.3 million. During which time they removed 1,510 hedgehogs. Which works out at £860 per hedgehog.

Work has not stopped; in fact some of the opponents to the cull have now been recruited to be part of the project's technical advisory group, including me. A big change came with the shift in focus – to begin with the project was measuring success by the number of dead hedgehogs – now the more sensible metric of living birds is used.

There is much to be done and very little money to do it with. Some of us, quietly, feel a little smug about one thing though. Right at the start of the business we suggested fencing as an option. This was dismissed as unworkable. It is now being seriously considered, and, as we also suggested, could be done in conjunction with surveying of the vulnerable areas and removing hedgehogs found where the birds are breeding.

Removing all the hedgehogs from the Uists is pretty much impossible and would only be feasible if a staggering amount of money was released to the now-renamed Uist Wader Research. It would have been easier if hedgehogs were not so popular – the extermination of rats and mink goes ahead with little fuss.

This is speciesism at work. As I have mentioned, we may think we approach these challenges with objectivity, but we are always making choices based on our own biases. If I was to have an uncompromising view of animal rights I would argue that the mink has as much right to life as the hedgehog. But I do not have that absolutist view. So, once we have accepted that there is a degree of subjectivity being applied, we need to look for it, account for it and incorporate it openly into our discussions.

If the decision to remove hedgehogs from the Uists had been taken with a harder mindset, well, there are techniques used in New Zealand that would have increased the chances of success.

But in New Zealand itself, at 268,000km^2, compared to the Uists at 700 km^2 – with a massive range of habitats including extensive suburbs where hedgehogs will congregate and be harder to kill, I think that removing all will be impossible. They will end up doing what will probably happen on the Uists: protecting the most valuable areas.

None of this is ever simple. The choices that have to be made, the compromises caused by cost and public opinion – all remind us that it would be so much better not to have to tidy up the mess of past generations. My hope is that we can learn from this, that we should stop making messes that will need attention in the future – to become, as my friend, author and philosopher Roman Krznaric puts it, 'better ancestors'. We also need to learn how to mix the objective and subjective viewpoints with some consistency. Above all, though, is my desire that we treat the world with more care, and recognise that the study of ecology should be approached with considerably more seriousness that it is at present.

Cane toad

There is a tale of wonder – as in 'I wonder what they were thinking' – that has many layers. It starts, as these stories tend to, with good intentions. The cane toad, a native of South and Central America, was introduced to Australia in 1935 by the Bureau of Sugar Experiment Stations in an attempt to control native cane beetles that were damaging sugar cane plantations. Crude insecticides had struggled to make an impact – not reaching the beetle larvae as they munched on the roots. So 102 toads were imported in 1935, allowed to breed, and their 62,000 toadlets were let loose in 1937.

There are now thought to be more than 200 million cane toads in Australia, spreading inland from the north-east coast. And with them comes a host of unintended consequences – not least their lack of efficacy as a biological control agent in the first place. They are a threat to native wildlife because they are poisonous, predatory, adaptive and competitive.

At all stages of their life, from eggs, through tadpoles, toadlets and adults, cane toads can kill predators if consumed. They have been linked to precipitous declines in the northern quoll, for example. This carnivorous marsupial almost completely disappeared from Kakadu National Park as the cane toads arrived. In an attempt to protect the quolls, those that remained were transferred to a toad-free island in 2003. By 2014 that population had substantially grown and were trained, via conditioned taste aversion, to stop attacking the toads. While this worked for a while, on their return to Kakadu, dingo predation of the quolls increased, and caused a localised extinction. This is another of those unintended consequences – it turns out that the quolls kept on the island lost their ability to recognise dingoes and cats as predators to be avoided, in the space of just 13 generations.

Cane toads themselves are really very unfussy as to what they will eat and basically if they can swallow it, they will.

They eat smaller amphibians, mammals and snakes. And the fact that they can do this in a wide variety of habitats – including urban and disturbed areas – alongside fast breeding, means they can rapidly colonise and, in the process, also outcompete local wildlife.

This has meant that the cane toad has been declared an enemy. The New South Wales government accepts that: 'The complete eradication of cane toads in New South Wales is not feasible, given their ability to thrive in a broad range of habitats, their capacity to reproduce in large numbers and their current widespread distribution.'

So that leaves the government with the aim of managing toads in national parks, but ensuring that no new populations are established, and trying to reduce their impact on biodiversity where they are already entrenched. There is no easy way to do this – there are no biological control agents (irony would be laden upon irony if they tried that and it went wrong as well) – and no other control methods that would not harm native wildlife.

The government advice is this: 'Cane toads must be collected and removed by hand. Traps and barrier fencing can be used to contain them but vary in effectiveness. According to recent research by the University of Sydney, refrigeration, followed by freezing, is the most efficient, effective and humane method of cane toad euthanasia.'

The take-home message about this species is that the best way to stop a cane toad invasion is simply not to start one.

CHAPTER 2

Residents

It is easy to assume that killing for conservation is reserved for invasive non-native species – and in most instances that is the case. However, there are exceptions, and where it is a native species that is causing trouble, the arguments are bound to be all the more complex.

One of my favourite places to go is Otmoor – the wet chequerboard landscape north-east of Oxford. It is a place of great history – a common marshland that was enclosed in the nineteenth century, fomenting the famous Otmoor Riots of 1829, and a wildlife haven that was threatened by the M40 extension but was saved by the action of local and national campaigners. As for its natural bounty, it draws in wildlife enthusiasts from miles around with the chance of seeing rare migratory wildfowl and waders.

It has been pointed out that what draws me to this wetland is tedious and predictable, like your favourite movie being *Casablanca* or favourite dog being a Labrador. It is the starlings. For years I have been one of the many who undertake the pilgrimage to see one of the wildlife spectacles of the world. And for me it is a short pilgrimage – just eight miles from my front door.

The walk from the car park can be interesting, depending on the water and mud levels, but eventually you will reach the hide and join the dedicated followers of dancing birds as we all wait hopefully for the fun to begin.

Most people understand the need for library-like conversations, hushed tones and cold toes. My aim is to identify someone who seems to know far more than me and stand near to them – my knowledge of wading birds is okay, but wildfowl I just keep forgetting. Most birders enjoy sharing knowledge. And it was on a cold January day that one of those experienced birders let slip something that really got me thinking.

It turned out he was a volunteer on the reserve. I mentioned one great conservation success story – the reintroduction of red kites which I loved to see and hear over my garden in East Oxford. I told of how I used to take the coach into London on a regular basis around 20 years ago, and as we went through the cutting in the Chilterns, I would scan the skies in hope of seeing one of these magnificent beasts. Their forked tails and size make them impossible to get wrong, which, for a very incompetent birder like myself, makes life a little easier. Now they are everywhere – so I need to find something else to augur a good day.

All positive, these amazing birds, I suggested. Not so, said my ephemeral guide. You see, the kites have a taste for lapwing chicks. The reserve is here to build up breeding numbers of waders, yet this is being hampered by predation from the usually scavenging kites. This presents the RSPB with something of a dilemma. How do they continue to support the lapwings without controlling the kites?

This was not the sort of discussion I had come to the reserve to hear – I wanted the simple delight that was about to unfold. Starlings. Maybe 40,000 that day would descend on the reed beds to roost and, if we were lucky, they would perform – murmurate – a wonderful dance in the sky, reminiscent of shoals of fish, twisting with impossible precision. Over the years murmurations have become increasingly popular – these magnificent displays are truly some of the most magical wildlife spectacles I have ever witnessed. I remember sitting on a boat off the coast of Dominica being taken out dolphin watching, and watching a murmuration evoked that same feeling of joy, awe and delight, accompanied by the sorts of noises usually heard at firework displays – oooohs and aaaahs!

I have had a long fascination with flocks of starlings. In 1985, during the first year of my degree at Leicester Polytechnic, now known as De Montfort University, I volunteered to help a PhD student with his research. He wanted to get estimates of the scale of flocks of starlings flying into the city from different points of the compass. I was allotted a park near where I lived, and found the ideal spot – a climbing frame. The weather was grey and there were no children to displace, so I got to the top, lay down to watch the sky, put on my Walkman headphones and returned to the Jethro Tull I had been listening to on the walk there … and lit my treasured and rare spliff. Possibly this display of dense clouds of dancing birds triggered the start of a deeper appreciation of the wonderful qualities of fieldwork. Though to be honest, I was never able to repeat the prog rock/hashish experience … normally too much else is going on.

Waiting for and watching wildlife is so deeply calming. And despite the blur of traffic in Leicester or the numbing of extremities on Otmoor, the rewards are nearly always worth the time. On this day, in early January, it was definitely worth it. First there was the movement in the distance right at the edge of unaided sight; like a summer swarm of flying ants, a flock of thousands of golden. On this day, in early January, it was for the night. A marsh harrier rose with unhurried ease from the reeds, binoculars showing the pale face turned towards my gaze, before she settled in to wait, just like I was. In the calm, snipe took the chance to charge across the open water for a new bit of reed, while the long-tailed tits, oblivious to the tension, gleaned.

At around 4.30 p.m. the first clouds of starlings arrived, their movements acting as a sign to others that here was a good place for the night. I noticed that the still water was acting as a perfect mirror with the sun setting behind us. The increasing numbers of birds were reflected as they headed for the reeds, then at the last minute decided to rear up. This happened for around 15 minutes, during which cold toes were forgotten. Not quite the ecstatic display it could have been – a stooping peregrine would have added a little extra to the aerial ballet. But really, I should not be so demanding. This was a magical afternoon.

As usual I waited until the bitter end, just in case there was another movement. Sometimes the starlings will decide on a different lodging for the night just as it gets too dark to photograph them. So it was really dark by the time I was walking the 2km back to the car park; luckily the puddles provided enough reflected light to keep me from veering into the boggy surrounds. I was thinking about the complexity of the problem that the RSPB faced. How would the public react if they started to kill kites to save lapwings?

Lapwings are amazing, but of the British wading birds, it is probably the curlew that has the greatest following – in part because of its mournful moorland cry – but also because of the work of just one person: Mary Colwell. She is to the curlew as I am to the hedgehog – she has written about them and campaigned on their behalf for years. And she has come to rather uncomfortable conclusions about what needs to be done to ensure their survival.

Talking of their survival is not hyperbole. In Ireland curlews have decreased by more than 90 per cent, in Wales by over 80 per cent and on average we have lost 60 per cent throughout England and Scotland since the 1980s. Curlew numbers are down thanks to industrial agriculture – so nothing particularly new there. But where this story gets really interesting is in two of the other drivers of curlew decline. One is increased woodland, which potentially sets a conflict with the rewilders who would like to see the moorlands return to their original form. And the other is predation.

Researching the relationships with predators led Mary to write a fascinating book, *Beak, Tooth and Claw*, a book that set some reviewers rather a-flutter. The challenge she presented was that where predators like mustelids and foxes are controlled (usually by gamekeepers intent on maintaining grouse moors), curlews benefit.

In 2022 I invited myself round for dinner – not quite as presumptively as that sounds, given that I used to work at the BBC with her husband, Julian Hector, who went on to run the world-famous Natural History Unit. We were due a good catch-up and it was a lovely evening, which ended with me falling asleep in the spare room, watched over by the most beautiful stained-glass curlew.

'What is winning?' This is how Mary started to answer my questions about the conflict between the various sides. 'We need creative collaboration – not "I'm right, you're wrong" bickering. We need to be willing to work through our distaste of other people's views.'

Julian came in with the food, joining in the conversation. 'The decisions that need to be taken are not easy. I have seen the sense of despair in the eyes of those at the conservation coalface.'

Certainly there has been despair from some at the conclusions that Mary reached – it can really stick in the craw, the idea of working with people with whom you have fundamental disagreements. Gamekeepers appear to kill hen harriers in their attempt to grow more grouse. This is illegal. But it also benefits curlews, who have one less pressure, although Mary is not condoning the illegal action of keepers. In her book she points out that it was a hatred of the raptors that caused them to disappear from the skies, and while stopping persecution will bring them back, she pointed out that 'it is only mutual respect, compromise and compassion that will keep them there'. This is an interesting take on the situation – though my understanding is that with predominantly conservative groups of people, if a law is enforced, then the problem is solved.

The threat of predators to her beloved curlews is only part of the problem and we soon ended up addressing another key threat – woodland. Now, while there might be some friction between the rewilders and the moor keepers, the real damage has been caused by industrial planting of Sitka spruce, often as part of carbon-offsetting plans. There is a requirement in many places for there to be a percentage of land given over to broadleaves, though this is what Julian dismissively described as 'just a veneer – a thin sheet of good wood hiding the ecologically rubbish trees behind'.

While Mary has the curlew in mind, the RSPB has a rather larger remit. They used to be quite coy about the need for occasional control of species that pose a threat to species they are trying to save. But they have more recently decided that being open with the reality of the challenges they face is a better option. Between 2012 and 2017, for example, they killed more than 8,000 animals in the name of conservation: 1,715 crows, 1,760 foxes and 1,734 roe deer, along with many others, and an unrecorded number of rabbits. I contacted them and was fortunate enough to get a Zoom call booked with three significant figures. Katie-jo Luxton is the director of global conservation, Paul Walton is head of species policy and George Campbell the operations director for the north of Scotland. With impeccable timing we arranged our online meeting to take place just as the RSPB was dealing with an upsurge in the latest outbreak of avian influenza and it was therefore a little shorter than we had originally planned. Rather than covering less, though, I think we just all talked and listened faster.

'I think you need to look at the work we do in the wider context,' Katie-jo began, briskly. 'The climate crisis, the collapse of ecosystems, changes in agricultural practice, are all having a massive impact on the ability of many species of bird to thrive. Then you throw something like this highly pathogenic avian flu outbreak into the mix. So with all of this going on, predation by vertebrates on already vulnerable populations can be the final nail in the coffin.'

This is at the heart of the issue – predation might not be the most serious threat to a population, but it is the one on which immediate action can be taken. Most of the other threats are just too big, and while it would be amazing if there was to be global resolve to tackle them, there is no point sitting on our hands and waiting for unicorns.

Paul Walton was the first to bring up the Uist saga. I was wondering whether there might have been any lingering concerns about this. 'The islands have arguably the most important breeding sites for waders in the whole of Europe,' he said. 'Birds are being hit hard by loss of food and climate change – across the wide range of their foraging sites, but there are only a few nesting areas, so the impact of predators is concentrated.'

'We never undertake vertebrate control lightly,' Katie-jo said. 'And anything we do has to go through an ethics advisory group, where it must meet a very strict set of criteria.'

She sent me the list. It has to be shown that:

- The seriousness of the problems has been established;
- Non-lethal measures have been assessed and not found to be practicable;
- Killing is an effective way of addressing the problem;
- Killing will not have an adverse impact on the conservation of the target or other non-target species.

George Campbell pointed out that even the control of a rodent infestation in an RSPB reserve building has to go through the same process. And Paul pointed out that 'our welfare standards are higher than the legal requirements'.

The RSPB is acutely aware of something that has the capacity to polarise: language. If you demonise and 'other' a species, it is much easier to call for lethal control. Paul explained, 'We always frame this as an anthropogenic problem, one that we humans have created. This is not about demonising non-native species.'

This is important – the discourse surrounding species being in the wrong place at the wrong time can swiftly turn quite unpleasant. It is not surprising that Goebbels used the metaphor of rats to depict the Jews being prepared for extermination. We should always be careful with how we communicate.

The RSPB has a hierarchy of action; prevention is the absolute priority. If a species does not become a 'pest' there is no need for control. Stopping a species becoming a pest can mean stopping it getting to a sensitive spot, such as an island. Or it can require a broader look at an altered ecosystem tipped out of balance. For example, when it comes to the kites around Otmoor, it seems that the solution for now may well be to do displacement feeding at wader breeding time, to draw the predators to different areas, away from the temptation of lapwing chicks.

If prevention fails, then rapid action is essential – not just to solve the problem the new species has created, but because if lethal control is being used for a shorter time, then fewer animals will end up being killed. If a cull is extended over many seasons, breeding will continue, and more animals will die.

One of the ethics advisors to the RSPB is Alick Simmons. Katie-jo said she would ask whether he would be willing to talk. I am still not used to this environment, where so many people are so cagey about being public. While I waited to hear back I did my own digging, and my interest in meeting him surged. He had just finished writing a book called *Treated Like Animals: Improving the Lives of the Creatures We Own, Eat and Use*. He is a vet with 35 years in public service, and also a naturalist.

Emails were exchanged and Alick agreed to meet. Finding the time was tricky, but in the end I managed to work him into a wonderfully busy day. After visiting him in Somerset in the morning, I was to head on to Devon and do an after-school hedgehog talk, followed by another for adults, just south of Dartmoor.

The village in which Alick lives could easily host a detective show; it's so perfectly peaceful that there must be darker dealings under the thatch. His house was busy – a new grandchild was sleeping on a parent, cooking was happening, and while he made coffee I was guided to the ideal office; not too big, but big enough to host a fascinating library. You can tell a lot about someone from the books they keep close, and his collection was encouragingly similar to mine, though a bit weighted towards ornithological monographs.

Armed with coffee, Alick began with an apology, the accumulation of family was due to a funeral later that day so he did not have as much time as he would have liked. Having seen something of his CV online I was just delighted he had managed to carve any time out at all – commitments include volunteering his time with the RSPB, and chairing the University Federation for Animal Welfare and the Humane Slaughter Association, among others.

We had already prepared some of the ground of what I wanted to talk about so he launched straight in with an extremely pertinent observation, which, embarrassingly, I had not really considered before, saying, 'If you accept that the majority of animals we interact with – from pets to hunting, the whole range, the vast majority of these experience suffering when exposed to physical or emotional stress. Most of these animals exhibit behaviour that suggests they are sentient. Yet we treat these animals in very different ways depending on the circumstances.

'Let's use a rat as an example,' he continued, and I got the feeling that this was not just the busy day that was letting ideas pour out so fast; this was how he was. 'Three imaginary rats to be precise. Here is my pet rat, treated with kindness and given veterinary care when ill. Then there is the rat I work on in the laboratory to better understand diseases in humans. And here is the rodenticide I put down to kill the rat that is eating my bird food. Three very different ways of treating the same animal – *Rattus norvegicus*. And also three very different legislative frameworks – the pet rat is protected under the Animal Welfare Act 2006, the laboratory rat is protected by the Animal (Scientific Procedures) Act of 1986 and as for the wild rat, well, there is minimal protection despite the many methods we have devised to kill them – few of which could be described as humane.'

I am not really interested in the control of pests, I explain.

'But you need to understand what I do with the RSPB,' he continued. 'And that is bound by legislation, as well as ethics. It is important that you are aware of the complexities. Different animals get treated differently, despite experiencing just the same amount of pain and distress. And I think it is important that this is accounted for. The purpose to which the animals are being put should be subordinate to their needs – societal attitudes are reflected in the law, details of the law then need to be enforced.

'The reason I bring up the rat is because it is very illustrative,' Alick said. 'Just look at the level of attention different animals get – there are 153 animal research laboratories in the country, which are attended to by about 20 inspectors. This means that each inspector has eight premises to check. But farms, well, there are around 80,000 farms and between 100 and 200 inspectors, depending on how you calculate it, meaning they could each have 800 farms to attend. And wildlife gets no routine inspection. Yet all of the animals concerned can suffer and we, as a society, have decided that some are worthy of more care than others.

'Take zebra fish, for example,' he continued. 'In a laboratory in Oxford where they use them as they provide a good model for neuroscience experiments, they have to account for every single one of these animals and there is trouble if they are mistreated in the eyes of the law. Yet look into a broiler farm – those chickens, probably more capable of suffering than the zebra fish, are routinely treated expendably, millions of them.'

I was seriously not expecting such an explosive agenda from the start. I mentioned that this was not far from what I hear animal rights activists talking about. But he was having none of that, saying, 'There is no point going with an absolutist, rigid approach to the complex areas we find ourselves in, we need to apply a utilitarian approach. When I am confronted with one of these complex issues I have to weigh up the harm and the benefit – and with the work of the RSPB this is often the potential harm to individuals versus the benefit to the population.'

Now this is something I can agree with from experience. We know that there will be some risk in attaching a tag to a hedgehog, for example. These tags are invaluable in giving us an insight into the habitats that hedgehogs use, and thus providing the sort of information that can help protect them. So the potential harm to an individual hedgehog – and I should add, this is very slight and happens rarely – is offset by the greater good of gathering information.

From his own experience, Alick described the work of ethics panels as 'places where decisions are made that are not necessarily correct, but have at least been thought about'. This is important; a lot of the mistreatment of animals – domestic and wild – comes from just not thinking about the implications of our actions.

His work with the RSPB is, as one would expect, meticulous. The RSPB kills thousands of wild animals every year in the name of conservation and they get attacked by both the gamekeepers, looking to score cheap points by trying to identify hypocrisy, and animal lovers, outraged at animals being harmed.

Many of these animals could be killed under what is known as a general licence. This is issued by Natural

England, and 'authorises activities that would otherwise be a criminal offence under Part 1 of the 1981 Act'. That act, the Wildlife and Countryside Act, is important legislation, but there are many concerns about the freedom the general licence brings.

So the RSPB operates not with a general licence but by gaining specific licences to act – and through this, very much higher standards are being applied. Alick explained how important it was that organisations like the RSPB hold themselves to account by allowing scrutiny.

'There was a real risk when light was shone on this,' Alick said, 'but the reality was mostly increased respect. Though every now and then someone with a grudge will "discover" the information again and make a fuss.'

Equally surprising was the point Alick made about the vital conservation tool of bird ringing – the process of putting small, unique metal rings on the legs of birds caught by a variety of means. Mist-netting is the most common – though I have chased fulmars around a small walled field and grabbed them by the neck to take to the ringer. Those fulmars were on North Ronaldsay, Orkney, and were in a small, walled field. They would have died if they had not been caught and eventually thrown into the wind, with a ring on their leg.

Those particular fulmars are a rare case. And any risk of increased mortality associated with catching and handling them was easily outweighed by their inevitable death, trapped as they were. But for other species, where the capture and the handling is just for the pursuit of knowledge, should we be concerned if it turns out there is an increase in mortality associated with the process?

'We need to look at this through the lens of utility again,' said Alick. 'Is the harm delivered to the individuals greater than the benefits gained from the increased knowledge? Is there a low rate of returns on the ringed birds because the process is deeply inefficient? Or is it because the process causes so many of them to die? Without proper scrutiny, this sort of question hangs in the air.'

If these questions are not asked then lessons are not learned – it was found, for example, that the tags put onto the legs of young cranes caused abrasions and injury, and so work was done to change the design. Alick feels that the three Rs of the animal-testing world should be applied to all research – even the apparently benign work for conservation: Refine, Reduce, Replace. If we are going to do the work, do it as well as possible; keep the number of animals used to a minimum; and finally, if we can do it without interfering with the animals, then that should be the goal.

'So, you remember I talked about the way the same animal is treated in different ways depending on who is in control of the situation,' Alick said. 'Well, there is much more to that we need to explore and address. Whatever your views on animal testing, the research world is held to the highest account when it comes to the treatment of animals.'

At this point I have to interject. My wife, the film-maker Zoe Broughton, spent some considerable time undercover in Europe's largest animal testing laboratory, Huntingdon Life Sciences, filming for Channel 4's flagship investigative programme, *Dispatches*. She was, by chance, given a job with the beagle puppies, the resulting programme was called *It's a Dog's Life*. The public furore was instant. The footage she managed to capture on her hidden camera, which could only film for one continuous hour a day, of the technicians shaking and thumping the puppies, reduced many to tears, and still does. Three of the technicians were arrested at dawn on the day after screening.

The two aspects Zoe captured that had a deeper impact were less obvious. One was that some of the technicians were fiddling the doses of the drugs being given to the puppies – so that all of the pain and cruelty were for nothing. This seriously upset the shareholders, and trading on the stock exchange was halted the day after broadcast as the share price plummeted. And the other was what Zoe reported – that the Home Office inspectors had visited, but they had stopped in the office and had tea with the staff and had not come in to see the puppies.

Alick paused, and then said, 'The research world is not perfect, but it is highly regulated and carefully scrutinised. Not all animal research is necessary and there is much I would stop. But, like many others, I owe my life to medicines that were developed using animals. Of course, it would be better if there were no animals used in research but until science comes up with robust alternatives useful in all circumstances, then some animals need to be used.'

As much as I find vivisection distasteful – my earliest political writing was a simple regurgitation of campaign material from Animal Aid as I argued against animal testing in a school project – I do appreciate the nuance. In fact one of the most disingenuous approaches of the ProTest (yes, clever name) movement that supported animal testing was to conflate the cutting-edge research done to try and combat disease with the toxicity testing that has to be done on every new chemical when it arrives in the country, whether it has been passed as safe elsewhere or not. Far more animals are killed for this toxicological work than are killed in pursuit of new drugs and treatments.

Again, though, both sides in the fight over this subject tend to take polarised views with such ferocity that the voices in the middle are drowned out.

Going back to the point he was making before I impertinently interrupted, Alick said, 'When you compare the research laboratories with the slaughterhouses, in terms of how they are held to account, they are in different universes. In a lab, there is a named individual who is held responsible for every animal and how it is treated and killed. In the slaughterhouse – well, the meat industry simply hides behind the regulator. The animals in the lab receiving a lethal injection after a life of ease and the animals in the abattoir hung by their ankles, stunned, if lucky, and then having their throats cut, after a journey in a truck of stress. They both have the capacity to feel, to suffer, to experience fear and pain.'

There is another comparison that bears attention: the RSPB is very carefully restricted in what it kills – by itself, by public opinion and by the choice of licences they receive. Compare this to the shooting world and there is yet another gulf of difference. What oversight does the shooting industry receive? Gamekeepers operate on private land, which they will frequently vigorously defend from those they deem to be trespassers. The only time that illegal acts get caught is when investigators manage to place hidden cameras. And as with Zoe in the animal lab, the response is always that the perpetrators are just bad apples – that the industry is self-regulating and clean.

'The efforts that go into killing to conserve the game birds are not regulated,' Alick continued. 'Some of the practices are undoubtedly deeply cruel – Larsen traps, for example, relying on the gregarious nature of corvids and in turn relying on one captive bird to lure others that are put to death in front of it. But what really seems to escape attention is that the process of the shoot itself is also unregulated.'

Strangely I had never thought of this point before, that the process of getting a licence to own a shotgun is entirely down to public safety. Are you a reasonable enough person to be allowed to own a weapon that can kill other people? There is an unhelpful knee-jerk reaction to that from me – straight out of Heller's *Catch-22* – for a person to want a licence to own a gun, they are, by definition, not reasonable enough to own a gun. Or maybe Groucho Marx is more appropriate: 'I would refuse to join any club that would have me as a member.'

There is no attempt to ascertain whether the person obtaining the licence is in any way competent at shooting.

And competence is important – the ability to kill rather than maim is something that should be at the heart of the industry. Yet anyone can turn up and blast away, should they have the money or connections. No one is accountable for the suffering this causes. No one has any idea how many animals are wounded and crash and slowly die.

'In Italy,' Alick said, 'to get a shooting licence you have to take a course. You have to prove you can identify the species that you might encounter and you have to show you can shoot accurately. This does not seem like a very big ask for an industry to copy in this country.'

'Animals should die as quickly as possible,' he continued. 'This is why I was opposed to the free-shooting of badgers. I was deputy chief vet at Defra at the time they were ramping up to kill hundreds of thousands of badgers in an attempt to control bovine tuberculosis. This was already a toxic debate, and one that the courts have been busy with – two judicial reviews, but contradictory results. For me, though, the key has to be whether it is humane. And shooting badgers at night is very difficult.

'If you shoot an animal, there are three outcomes an observer can record. Shot and killed; shot, wounded, caught, killed; shot, missed or wounded and runs. And obviously most data shows the first two outcomes as the others are not found – yet this non-retrieval rate is important to know.'

In 2014, the independent panel of experts in animal welfare, veterinary pathology and badger ecology, which was appointed by Defra to oversee the humaneness of the badger cull, made this conclusion. Between 7.4% and 22.85% of badgers that were shot at were still alive after 5 minutes and therefore were at risk of experiencing marked pain…'

This has to be added to the equation or spreadsheet – we have to be aware of how much suffering is caused. If you have taken the normative (deontological) view that all killing is wrong, then this detail is irrelevant, but I am increasingly finding that while my heart may demand one thing, my utilitarian head is taking me in a different direction. If the animal is rendered instantly unconscious so that it can die with little distress, then that is better than suffering for 5 minutes or more.

Alick Simmons was proving to be a more influential thinker on this than I had imagined. Soon after our interview I received a pre-publication proof of his book, *Treated Like Animals*. It is full of the important questions that I had often not even thought to ask. He manages to maintain scientific detachment while ensuring that those involved in the harming of animals are held to account.

Applying a bit of scientific detachment to another bird-related conundrum seems in order. There is big business in shooting and, as Alick had already explained, this is an industry that could seriously do with some additional controls.

There is a strange and labyrinthine intellectual and political exercise exerted to keep the shooting of pheasants and partridges legal. Early in 2022, wildlife law was changed to extend the permission gamekeepers had to kill crows and jackdaws in order to 'protect' target animals. In a lengthy tweet, George Monbiot takes us into the Pythonesque reasoning.

To justify gamekeepers' killing of corvids, pheasants and partridges need to be classified as livestock. But as you are not allowed to shoot livestock for sport, when they are released, they are transformed into wildlife. As you are not allowed to round up wildlife and trap them, they become livestock at the end of the season.

Now, if a pheasant flies into a car windscreen, the person who reared and released the bird is not liable because at that moment it is wildlife. Though if it survives the crash and you round it up to create more pheasants, it is livestock again. And only as a keeper of livestock can you claim tax breaks and subsidies.

George goes on to wonder who wrote these laws, and of course, it is the very people who own the shooting estates.

The sheer volume of birds released is astounding: 47 million pheasants are bred and released each year. The biomass of pheasants and partridges released each year is greater than the entire biomass of native birds.

Just let that sink in for a moment.

And the consequences of this? An utterly distorted rural fauna. When I was researching my book *The Beauty in the Beast*, I met an adder expert in Norfolk. He was not a great fan of people. But he loved his snakes and took me out into the Holt Lowes nature reserve to find some. As we walked he talked about the impact of pheasants on his beloved reptiles. How in areas with pheasant shoots all the baby adders get eaten out of the hedgerow by the omnivorous birds – while they will be fed grain, that does not stop them eating other things. He was upset and angry but I did not know how serious an issue it was.

The impact of these released birds is greater than just their predation of our native wildlife. Remember, there is a greater amount of meat out there on just these two species than on all the rest of our wild birds. And these birds die – some killed outright by those having fun with guns, some wounded, who flap on to die in pain elsewhere. Due to them not being immortal, they all end up dying, which provides a great deal of food for someone else. Great, you might think, they have a utility. But again, this is ecology and therefore this is not straightforward.

The pheasant and partridge carcasses strewn across the land unbalance further an already enormously degraded ecosystem. Those who benefit are the scavengers, the foxes and corvids, the very animals that are being killed to protect the target birds in the first place.

George concluded his tweet with this delightful sign-off: 'Is this country a democracy, or a semi-feudal barony? It's a gigantic, state-sanctioned cosplay, simultaneously deadly serious and utterly risible.'

Of our resident species, it is not just foxes and corvids who end up being killed to conserve the shooters' privilege. Now no one is going to admit to doing it, as it is illegal, but there is a strange pattern of disappearance of hen harriers in and around shooting estates. These majestic birds – sky dancers – have quite a following, which gathers once a year to share knowledge and celebrate a solidarity of defiance. I was not planning on heading to the Hen Harrier Day in 2022. In fact, despite repeated invitations, I had never been to one of these Chris Packham – organised events. Not out of any objection to the work they do – highlighting wildlife crime that takes place in and around grouse moors – but because it happens in the summer holidays when I am juggling kids, and also because it always seems to rain.

But in 2022 the event coincided with my daughter, Raine, as she now wants to be known, needing to get to Edge Hill University, north of Liverpool, and the event took place in the grounds of Adlington Hall in Cheshire, not too far from my mother in Birkenhead. So on a lovely dry Sunday we headed north and arrived to join a gathering crowd of more than 700 people.

We walked up to the stage to see the line-up of speakers, who were really quite inspiring: Natalie Bennet from the House of Lords, Sheffield MP Olivia Blake, and of course the stars of the Wild Justice campaign, Chris Packham, Ruth Tingay and Mark Avery. The hard work of organising was being done by the very twinkly Lucy Lapwing, one of a growing number of young, enthusiastic and very good communicators. I had not met her in person before, so leapt in with my usual reticence and pointed out that the speakers looked great, but there was an absence of anyone talking about hedgehogs … and that if anyone dropped out I would be happy to step up.

Timing is everything. 'Someone has just dropped out,' Lucy said. 'Did you mean it? Would you be able to do something?'

Charmed that she had framed this as a question, I leapt at the chance, and then had that moment of realising I had to find a quiet space to prepare. Inevitably, the rain started, but the weather app on my phone suggested it would only be brief and light. Raine and I took refuge in the gazebo that doubled as the 'green room', though Raine did spend quite a while queuing for pizza, wearing my waterproof. It is not inconceivable that she timed this so that she would miss my moment on stage (and thank you to the audience who stayed, despite the rain – there is a move to offer the event to areas of the country experiencing drought, as it seems the best way to guarantee a downpour).

I only had 10 minutes; Lucy in the wings was pointing at her watch as I drifted into tangents. My message was simple, it boils down to finding ways to fall in love with nature. Liking is okay, but not enough. 'We will not fight to save what we do not love,' said Stephen Jay Gould, something I try and get into every talk I give.

I also made people laugh, by reminding the ornithologists that hedgehogs remain more popular with the British public than birds, which, given the seriousness of the rest of the speakers and the torrents of rain, was necessary. I was eventually whisked off stage by a very tolerant Lapwing and went to find my slice of pizza (Raine got three!).

While I was there I noticed a couple of people who rather stood out. They were bigger, in a well-muscled sort of way, quiet and very interested in everything that was going on. My suspicions were raised. In many of the campaigns that I have been involved with there has always been a steady stream of infiltrators – people employed by either the state or private companies to spy on peace campaigners and environmental protesters. So much of a problem has it become that a judge-led inquiry into undercover policing was set up in 2015 looking into this, in particular where these undercover operators entered into long-term sexual relationships with campaigners.

I tried chatting to one of them; he was very uncommunicative. But a horrible feeling began to grow as it dawned on me, they were not here as spies, but as protection. For Chris Packham. When I got that confirmed my heart sank. I have had the pleasure of visiting Chris at his New Forest home on a few occasions. When I was last there, filming with his step-daughter Megan McCubbin, workers were just finishing rebuilding his gates, which had been burnt down when people drove a car right up to them and set it alight. All caught on camera. All very professionally done – this was not just someone randomly sloshing petrol over the place. There was an explosion.

Chris and Megan told me about the dead animals that have been left on their gate posts, about the threats to them – threats that the police have had to advise them to treat seriously. They have been on self-defence and de-escalation trainings, learning how to manage aggressive encounters. I think Chris is more worried about the threats to kill his beloved poodles, Sid and Nancy.

When I mentioned this to Raine, her response was, I think, pretty representative of most people in this country: 'I don't understand, he is just so lovely.'

The threats of violence can only be seen as an indication of the fear Chris causes in those who are keen to continue hunting. He has got them so worried that they are willing to resort to such extremes.

The nature of these heavily polarised debates has really deteriorated. When a side willingly employs tactics that are designed to terrorise, you need to consider what sort of group it is. When the Earth Liberation Front destroyed property, being very careful to avoid any injury to people, they were labelled as terrorists. What does that make those that threaten to kill, and destroy property in a manner that really does risk human life?

Before I left the Hen Harrier Day, I said my goodbyes and mentioned to Chris that the following week I was going to be not far from his home, taking photographs at the New Forest Show. This is another component of my portfolio career, as I understand freelancers now define their livelihoods. I would be working for Guy Reece, who is a superb photographer of people – his company Striking Faces gets employed by festivals to collect amazing images of the crowds enjoying themselves. I was there as backup, as the event is so large he cannot cover it all.

There was another reason for wanting to go, and that was access – access to the sorts of people I do not normally meet. The hunters and the gamekeepers, for example. I mentioned this to Chris and I could feel his hackles rise. I had not realised, but it is from among the people who were to be at the New Forest Show that the death threats have emerged. It is not impossible, he said, that the very people who burnt his gates would be there, enjoying their time in the limelight.

That certainly left me feeling uneasy.

The New Forest Show was quite an experience. Lots of it is dedicated to consumerism – shopping for your country lifestyle, whether your SUV ever saw mud or not. I had expected to feel intimidated, entering the community of those who had given Chris and Megan such a hard time, and I had expected to feel anger as well. But I had a camera and a job, so I got to work.

It was fascinating – the hunts of various forms were there as if they did not have a care in the world. Legislation might have interfered with their 'sport' but from the way they carried themselves, it was as if those laws meant very little indeed. In fact undercover work from animal welfare organisations has shown as much, with the emphasis on ensuring you don't get caught hunting rather than hunting in line with the legislation.

The days I was there were hot and there was a lot to see and photograph: competitive woodchopping, terrier racing, LARPing traction engine fans and 'land girls', competitive blacksmithing and horses – lots of horses.

But by the time I found myself among the gamekeepers my initial confidence was beginning to wane. I am not sure what it is about that world that affects me so, but it does. As much as I love the idea of open dialogue, of trying to talk through issues, to understand the point of view of the person I disagree with, in this situation I again feel the rise of anger. Obviously I could not go in and start accusing people of threatening violence to Chris, and I could not go in there angry. I needed to find a way of understanding, because I would never be able to get through this if I did not understand.

This is not the first time I have found myself in such a situation. For three years I attended *Countryfile* Live – a monstrous imagining of the long-running BBC countryside programme. For two of those years I had a stall; I sacrificed the family holiday trip to the wonderful Green Gathering festival, where lots of my friends drank and danced and made music into the early hours. I missed out on the sauna, a brilliantly reformatted horse box. I missed seeing my dear friends Theo and Shannon, from the inspirational political folk band, Seize the Day, and their daughter, Rosa.

I missed all of that because I got the chance to spend four days talking to hundreds and hundreds of people about hedgehogs. I had a team of volunteers who helped keep things running, helped children make clay hedgehogs while I bent the ears of their parents. It was exhausting but worthwhile.

Each day I would give myself a 30-minute break to see other parts of the event. Mostly it was hideous. Two warehouses were transported to the scene, one filled with potpourri and sub-garden-centre tat, and the other stuffed to the gunnels with variations on dead pig and gin.

There was also the vast area given over to the country sports proponents (our wildlife space was carefully hidden away). I tried, I really tried, to go and talk to the gamekeepers. But I realised that I carried an anger that would not make for constructive communication. To be honest, I felt like I would probably shout or cry. There was an air of the boarding-school bully about the entire tentful of them.

So with this experience in mind, for now I retreated.

Capercaillie and the pine marten

I have never met a capercaillie, but I would love to – they are magnificent, the largest of the grouse, the size of a turkey. The male's lekking behaviour makes him one of the icons of Scottish wildlife. He asserts his dominance with a display – with play, as *lek* means 'play' – that sounds deeply unlike a bird. The quality of the lekking males' dance and song, a combination of clicks and clatters, wingbeats and wheezes, determines which of them will have the chance to mate.

The population of capercaillie is declining fast, and there are just over 500 left in Scotland, meaning they are at a very real risk of extinction (again). They have already been wiped out once by the mid eighteenth century thanks to loss of woodland habitat, especially mature Scots pine. They were reintroduced from Sweden in the nineteenth century to meet the desires of hunting folk, but have suffered a particular decline in recent years.

I *have* met a few pine martens. They are quite the cutest of the mustelids. Yes, they are brilliant predators, but they do so with charm. Admittedly the ones I have seen have been in wildlife centres, so I have not seen them on the hunt. But they are beautiful and I suggest you go and have a quick look in your wildlife guides, or online, if you are unfamiliar with them. They have dark brown fur with a creamy-white bib and can be up to the size of a small cat.

Watching a pine marten's liquid form move around the enclosure at Rewilding Coombeshead in Devon – part of the work of the indefatigable 'beaver breeder from Broadwoodwidger', Derek Gow – was mesmerising. These beautiful animals have also suffered severe population declines. Once ubiquitous throughout the UK, there are now an estimated 3,700 in Scotland. They were subject to the attentions of farmers and gamekeepers, who disliked their potential to harvest avian crops.

Pine martens became fully protected in 1988, and this has allowed their numbers to slowly increase. Which is a wonderful thing. But one of the targets of their appetites are the eggs and chicks of the capercaillie. And in a review by NatureScot in 2021, they were identified as one of the obstacles to the capercaillie growth.

What is to be done? The problems the bird faced used to be from corvids and foxes. And conservationists would intervene to increase the breeding success. By intervene, I mean, as you can probably guess, kill. But now that we see that capercaillie breeding success is also negatively correlated with pine marten activity, conservationists have a quandary. The martens are protected, and loved by a far wider audience than corvids and foxes were.

One Swedish study, back in 1988, showed how the culling of foxes and martens was associated with an increase in capercaillie breeding success. And the NatureScot report concluded, 'Management initiatives aimed at removing predators can produce immediate results, but they pose practical and societal challenges.' Another option is diversionary feeding – while the grouse are nesting, extra food is supplied at a distance to entice the predators away.

Oh, and just to keep things complicated, badger numbers have also been increasing in capercaillie-favoured forests, and they too have a fancy for capercaillie eggs. Now, these predators are not the only threats these birds face. Disturbance from humans, habitat loss and fragmentation by fences, and the impacts of a more erratic climate are all also known impediments to a healthy capercaillie population.

This story highlights the complex nature of wildlife management. These are two species that have received considerable investment to reestablish them in the wild. The arguments about letting nature take her course are redundant as we have already intervened, considerably. If it comes down to choosing to kill one to save the other, would you say yes? And perhaps more importantly, could you do it?

CHAPTER 3

Squirrels

Who are the aliens? Science fiction has taught us that they will be obviously very different; large heads, or tripods for bodies, or green, or with a luminous finger. And as for wildlife: surely if a species is in the wrong place it will stand out. Just look at those beautiful ring-necked parakeets. I still get a thrill from the almost tropical delight they bring to London parks. They are stand-out different, with green feathers as loud as their call.

There are records of them in the United Kingdom back to the mid nineteenth century, but it was not until the late 1990s that they became obviously established. Researchers have found that most of them have their origins in Pakistan and north India (part of their native range), and that while there are some lovely stories of the feral birds' origins (like the one about Jimi Hendrix releasing a pair in London's Carnaby Street), the reality is more mundane. There have been many occasions on which pet birds have been released or escaped, and survived in the wild. And now, they are here to stay, it seems. They have spread up the country; I often see them in Oxford and they have made themselves at home as far north as the Scottish Borders.

These aliens are beautiful, yes, but they do also have an impact on our native wildlife. They nest in cavities in trees, and can outcompete woodpeckers and nuthatches. They can also swamp garden bird feeders, to the annoyance of generous suppliers of nuts. I have spoken with ecologists who wistfully comment on the moment when they could have been brought under control – but now, with around 50,000 of them scattered through the suburbs, we are too late. And there is the additional problem: people love them.

On my way to a photo job in London I walked through St James's Park, and soon came across a crowd of people. I had been in my own world, listening to the radio probably, so it took me a moment to work out what was happening. People had their arms outstretched, fingers baited with half apples or nuts. And the objects of their adoration were waiting and then fluttering in for the offering. These parakeets had managed to tame people into their service.

Looking back at the photographs I took – in particular of the kids with these gorgeous green birds on their hands or heads (permission sought from parents, photos sent on to them as well) – the joy of this moment of connection with wildlife will surely be the reason why the parakeets are safe from conservationists. One of the children said, with utter glee, that this was the best bit of their holiday – so good was it that this was a return visit after their parents had spent no small fortune on taking them to a show, museums, meals out and Madame Tussauds.

While not as embedded with the public as the parakeet, another exotic bird was the subject of a UK government eradication campaign back in the 2005. This was the ruddy duck, an import from North America. First seen in the wild in the 1950s, having escaped from captive wildfowl collections, there was an estimated population of 6,000 ruddy ducks when the cull began.

The purpose of the cull was to help in the conservation of their 'cousin', the white-headed duck. Both are from the genus of 'stifftail' ducks, they occupy similar ecological niches, and – here is the problem – they are able to successfully breed together.

White-headed ducks are a European conservation success story. In 1977, the Spanish population was down to just 22 birds when a ban on hunting them was introduced alongside active measures to protect breeding sites. Now there are around 2,500 birds, which is a success for more than just the species. So iconic has this project become that it has helped reinforce protection of their wetland home, as well as control hunting. If it was to vanish, a key argument for maintaining this ecosystem would also be lost.

Ruddy ducks from the United Kingdom were found to have spread onto the continent. And on reaching Spain, they started to do what ducks will do, leading to the real risk of a dilution to the point of extinction of the white-headed ducks. Spanish authorities worked hard to remove the ruddy ducks, but they were replenished by more birds from the United Kingdom, so the decision was taken, in the name of the conservation of birds in another country, to kill the UK ducks.

By March 2008 ruddy ducks were reduced to around 400 in the UK, by 2019 the number was down to less than a hundred. This has been at quite a considerable cost – estimated to be around £3.3 million – and the remaining few do, and should, cause concern. If they are allowed to re-establish themselves, not only will the money have been wasted, but so will the lives of the many thousands of ducks that were killed to get to this point.

This did not happen without outrage, of course, as culling efforts were disrupted by protesters. Birders have been reticent about reporting sightings, knowing that this will result in the ruddy ducks being killed.

Those against the cull have made the observation that this was a politically motivated action. For example, the campaigners at Animal Aid argue that Spain, backed into a corner over its failure to protect important habitats for white-headed ducks from intensive agriculture, hit back with a call for action on ruddy ducks.

Campaigners at Animal Aid made their position clear:

We find this violent eradication programme unacceptable, especially when a humane alternative is available. It is sadly now too late to save the majority of ruddy ducks in Britain, but, as wildlife lovers, we need to come together to make sure no other 'non-native' species suffer this same fate. We must be vocal about our support and challenge those spreading misinformation.

They ask anyone who sees a ruddy duck to keep quiet, saying, 'Even bird groups will tell the authorities and those birds may be killed.'

My friends in the animal rights world are clear about this sort of conservation – saying it is wrong. They have taken, and thank you to the philosophers who have shared a coffee with me to help get me to understand, the normative or deontological path – if there was a runaway tram, they would not pull the lever.

Not all aliens arouse such strong feelings. Red-necked wallabies, some released by exotic animal collections at the start of the Second World War, are clinging to this country by their dextrous paws, managing to eke out a living on the Isle of Man. I remember heading up to the Peak District with some students to practice survey techniques for our expedition to Morocco in search of an extinct leopard – but using wallabies instead. There were still some there in the very early 1990s, but all are gone now I fear. The survey technique was questionnaire-based – there are very few similarities between an extinct leopard and a small kangaroo – though the wallaby had the advantage of actually existing. If anyone does recruit you to go looking for an extinct leopard, be prepared for disappointment and dramatic weight loss.

Part of the reason for the lack of worry about wallabies is that they really were only just managing to survive. If there were suddenly hordes of them sweeping across the bleak peaks, maybe concern would have been raised. And they also have the advantage of simply being very endearing.

But if we are to confront alien species with lethal force, surely we must do so with some consistency. These decisions are too important to base on whim, or levels of cuteness. We need to know what we are trying to achieve. Are we choosing a point in history that we hope to get our ecosystem back to, like a computer system reset? Do we take it back to the blank canvas and start again? Or do we draw a line? And if we are to draw that line, where should it be?

For example, consider *Lumbricus terrestris*, the common earthworm; the largest of the worms native to Britain. This species is of massive ecological importance – it is a detritivore, redistributing nutrients in the soil, creating a more porous substrate to enable water and nutrients to move. Without this worm, our ecosystem would be seriously impacted.

But what about *Lumbricus terrestris* in the Americas? There were none until colonialists inadvertently brought them over in the eighteenth century in the soil surrounding the bulbs of plants being imported. And while the plants themselves have created the more obvious transformation, the worms have utterly transformed vast areas of many states.

There were worms in the Americas before they arrived, but the northern states had none; they had been wiped out by the previous Ice Age. This had resulted in massive accumulations of dead plant matter from which the forests bloomed. The arrival of the earthworm saw their engineering skills transform the land, pulling the leaves down and deep into the soil, away from the saplings that were attempting to grow.

There is no way these earthworms could be removed from the Americas. Their arrival has dramatically altered the ecosystem. So it is clearly impossible to recreate a 'pure' land. And once that has been accepted, the fallacy of polarised thinking should be obvious.

In Britain we have a series of 'invasions' to consider. Leaving aside the land bridge to the continent, the Mesolithic hunter-gatherers, the early farmers of the Neolithic, the Bronze Age and onwards, all of whom altered the postglacial ecosystems, let's start with the Romans. With them came the brown hare, which is assumed by most to be an iconic native. Domestic cats and fallow deer also arrived back then.

The Normans – well, they have a lot to answer for in the way Britain is governed, but looking at their contribution to our wildlife, the most significant would be the rabbit. Of all the terrestrial mammals living in the wild in the Britain, 11 are not native. So are we to consider brown hares, rabbits and fallow deer as aliens we need to control? I would suggest that would be nonsense. Moving forward in time, where do we draw the line? And is that even a sensible way of approaching the issue of new arrivals? If we do not have a line, there is the risk that every decision has to reinvent all the arguments again – after all, taking a case-by-case approach is much more time-consuming. But I do believe we have to take each case as it comes. Wallabies and parakeets are welcome as far as I am concerned. The blindingly obvious reality is that there is no line to which we can return this country. There never has been and there never will be. Ecosystems change, and it is only us, wise apes, who think that things should remain a particular way. Which does not mean we should not intervene, just that there is no point trying to recreate some mythic past.

But what of one of the most divisive non-native mammals we have in Britain, the grey squirrel? The first records of *Sciurus carolinensis* are from 1876 when a pair, kept in a cage at Henbury Park near Macclesfield, Cheshire, was released after the owners, who had brought them from America as an ornamental species, became bored with their acquisition. This pair, and their offspring, thrived and soon were noted as a local pest. Perhaps more important in the establishment of this species were the bigger releases in Woburn Park in 1890, Richmond Park in 1902 and Regent's Park between 1905 and 1907. The last known import was in 1929, and by 1937 a law had been passed banning their importation and release due to their damage to the forestry industry.

The pest value of greys is also a conservation concern. They strip bark, creating open wounds that can weaken, stress and even kill trees. The need for healthy broadleaf woodland is well understood, whether as a carbon store, or as a key starting point for increasing biodiversity and bioabundance. We know that trees are also important for our own well-being. If left to their own devices, between April and September, juvenile grey squirrels will target young broadleaved trees of between 10 and 40 years old, and they will repeat this every year unless 'managed'. This level of damage certainly reduces the amount of commercial planting of broadleaves as it becomes uneconomic. In particular the greys concentrate their attentions on oak, beech, hornbeam and sweet chestnut, largely leaving lime, horse chestnut and wild cherry.

The much-missed Derek Yalden points out that the tale of the grey squirrel in the United Kingdom coincides with the beginning of ecological study. His fascinating book, *The History of British Mammals*, is essential reading. By 1944 it became evident that where greys had been established for around 20 years, native red squirrels had disappeared. Over the next 40 years the greys' spread was charted and the effects noted. They simply brushed the reds from the map.

Interestingly, this is not the first time that the red squirrel has suffered such a fate. While it now has its stronghold in Scotland, it was pretty much exterminated there in the eighteenth century. This was at a time of intensive deforestation, and before the arrival of plantations. But it was not just inadvertent; reds were also targeted as a potential threat to the regrowth of trees. Some of the subsequent return of red squirrels was deliberate – and founded not from the United Kingdom; Scandinavian stock was imported at the end of that century.

It seems we are never quite clear on whether we love them or not, because reds were then identified as threats to new conifer plantations. In 1903, the very year Beatrix Potter's Squirrel Nutkin was doing for the species what Mrs Tiggy-Winkle did for hedgehogs, the Highland Squirrel Club was formed – not to embrace the reds, but to kill them. In just three years, club members slaughtered over 85,000 of them, to the point that sympathetic conservationists and landowners had to restock them, again from Scandinavia.

The red squirrel is one of the most charismatic mammals we have in the United Kingdom – delicate and dextrous paws, tufty ears, ability to tolerate proximity to humans . . . there is a lot to love about them. Not least as food – the incomparable wildlife journalist Patrick Barkham reported that in the early nineteenth century around 20,000 red squirrels were sold in London meat markets each year.

But they are also, now, one of our most endangered mammals. The British population for red squirrels was estimated to be around 3.5 million but in the last 60 years or so it has plummeted to around 140,000 – a decline of 96 per cent. The damage is not evenly spread; they have been almost entirely eradicated from England, with a few pockets in, for example, Merseyside, the Isle of Wight and Northumberland holding out. The bulk of what remains live in Scotland.

This is despite having some very high-profile advocates. In 2009 the then Prince of Wales was present at the launch of the Red Squirrel Survival Trust. Rather misguidedly, he argued that, 'I cannot think of a better mascot for our country than the red squirrel', when clearly the hedgehog is the perfect mascot for our country, always beating Tufty in votes for our nature icon.

In 2011 the Red Squirrel Conservation Project was set up. Again the Prince of Wales was present, saying, 'My dream is that red squirrels might thrive in the UK and it is here in the north of England that perhaps we can dare to think it might be a reality, thanks to people like yourselves.' He added, 'My great ambition is to have one in the house ... sitting on the breakfast table and on my shoulder!'

Now I am not sure that conservation should use the metric of 'who shared my muesli' as a measure of success. But at least there is persistent energy behind the campaign. Much of the focus of these campaigns has been on trying to halt the progress of grey squirrels. But on the mainland this is a task that makes Sisyphus seem like he got off lightly. If we are to kill to conserve, it seems reasonable to argue that there should be a goal, an end point.

This is where the conservation of red squirrels on Anglesey can teach us a lesson. Ynys Môn, as Anglesey is know in Welsh, is an island with a special place in my heart. Not, as might be expected, from a connection with nature, but from being the place where I learned to drive. I would have been 13, and one summer holiday I was invited over to visit a school friend who lived on the island. His parents had a farm up on the north coast, where a rather casual attitude to health and safety made it a very entertaining adventure. I got to fire a shotgun for the first time, just for the joy of the bang, not *at* anything. And then there was the car, an old Triumph Herald. After I had been given a few rudimentary introductions into the mechanics of making it move, off we went, bumping over a large field.

The first couple of 'pings' were largely ignored, in part due to the noise of the car, but they kept coming until my friend, Tim, worked out what was happening. His older brother was by the front door of the house with his air rifle, using us as target practice.

The fact that I had not been back to Anglesey for 40 years was not due to this experience, just a lack of opportunity. But I had a real desire to revisit the island because it had become quite famous in 2014 as an example of successful control of an invasive species. So famous that Prince Charles talked about it at the launch of the Red Squirrel Conservation Project, saying, 'I have been lucky enough to follow the world-class work on Anglesey of Craig Shuttleworth who really is extraordinary in terms of his energy and commitment.'

Clearly Craig was my man. Though I was surprised, when I reached out to him, that he was not exactly effusive. There were a couple of slowly answered emails, followed by an arrangement for a telephone call. In my 'usual' world I find most people tend to be very open, willing to talk about their work, sometimes to a point where even I feel enough is enough. But Craig was reticent. This confused me; he was at the heart of an internationally recognised conservation programme that had done something many fail to do: succeed.

In retrospect the conversation was more like a job interview, with Craig checking me out, wondering whether I was worth the investment of time and energy. Cagey sounds too negative, but he was cautious. And as we talked it became clear why.

In 2018 Craig had done an interview with the BBC, talking about his work, and that had resulted in an avalanche of abuse. Social media did what it can do so well – erase nuance – and had turned a discussion about science and ecology into an onslaught. His home address was posted online, along with photographs of his children. It was an awful time.

His crime? Killing squirrels. Grey squirrels, to be precise.

We agreed to meet. Fortunately I have an old dancing friend who lives on Anglesey; Dr Vivian Shaw is a remarkable force of nature, merging ancient and modern in her research and practice, as she studies both anatomy and acupuncture. Unfortunately she was away at a long-awaited spa session with her wife on the night I needed a bed in order to be up and bouncy for Craig first thing in the morning. Fortunately, hospitality did not need a host, and I was told where a key had been hidden, and invited to make myself at home in her gorgeous house right by the Menai Straits.

My sat nav suggested a fairly direct route, but as I looked at the map I saw that actually I could quite reasonably drive through Snowdonia. Early March 2022 gifted me a stunning day. I listened to Helen Macdonald's sparkling essays in the audiobook of *Vesper Flight*, and felt growing excitement as I began the drive. Eventually the hedgerows of the Cheshire plain gave way to the granite of mountains. This was a real adventure. It had been two years since I had last felt such freedom to explore. And I had no time constraints, as no one was waiting for me to arrive.

Occasional stops to marvel at the geophony of waterfalls, the cronking of ravens and the wonder of an all-day vegetarian breakfast kept me elated as I headed on towards the Menai Straits and the island. Driving over Telford's 1826 suspension bridge and risking a glance at the turbulent sea below, I was happy to think that at least squirrels would not attempt crossing by water.

I almost messed up the next morning, as I had been distracted by the signpost beside the Waitrose supermarket car park which said *Coedwig gwiwer goch*: 'red squirrel woodland'. I pulled in there to wait, having failed to read Craig's email correctly, which directed me to an Asda car park back on the mainland.

Fortunately Craig was tolerant of my understandable, bumbling mistake and he greeted me warmly. Along with his well-worn wellies, combat trousers and battered fleece, his strong handshake reminded me of all the time my lily-white hands have spent inside at a computer.

In these days of easy access to history, I do wonder how much of these initial meetings – supposed to be the decisive factor in how things go – are catalysed by Google. Of course I had looked Craig up, and found the stories of his conservation success. I would not have come without doing so, but perhaps the ease of meeting was because he had checked me out too? I must find out what people see when they start searching, though it is a bit late now, I suppose!

I was asking a massive favour of Craig. Like myself he is self-employed, so by taking his time I am taking him away from work. But again, like me, he seems to have many different projects on the go at once, and appears to do many of them simply for the love of it. We had a little small talk, but it was clear he had an eye on the time and had other work to do.

'Before we start looking for squirrels,' he said, 'I have to go and feed the pine martens. Do you want to come along?'

That is not an offer I can imagine anyone in their right mind turning down. Despite their taste for capercaillie eggs, pine martens are incredibly special, and there has been some evidence that their presence can reduce grey squirrel numbers. So work to encourage them back has to be good news, after centuries of persecution.

I followed Craig's van as he weaved out into the countryside. Fortunately the roads were not too busy and I did not lose him – I was still feeling embarrassed about having gone to the wrong meeting point and did not want to compound this with a pitiful 'where are you' phone call. We pulled in near a forestry path that wound up a tree-covered hill just outside Bethesda.

Lockdown has done me no favours – as with many, I am sure, I have found that some of my clothes had shrunk. But I had not realised how unfit I had become. The journey up was okay, we walked and talked, and while I found it easier if I left the talking mostly to Craig, I could keep up. Then it was a scramble down between the rows of conifer on the other side, where I was able to speak more. It was only on the return journey, back up, that I found myself reeling a little. This was a clear message to me from my body – get fitter or forever be the one leaning on a moss-cushioned fallen trunk, wondering if I was really about to pass out.

In my defence, Craig was younger than me (by a couple of years, to be honest) and had been doing this all through lockdown and in fact pretty much all of his life. But even so, I was chagrined by the experience. And while we were on a mission to feed pine martens, we were not going to be seeing any, so perhaps that did diminish my reserves.

The feeding Craig was doing involved placing eggs from the supermarket into special boxes attached to trees. Of course, for the martens to thrive, they will need to be able to make it on their own, but these are being given a helping hand. They were released into this woodland in June 2020, having been bred at a number of different wildlife centres in Wales, Kent and the New Forest.

I was rather pleased I made it back to the car in one piece, and the last leg of the journey, downhill, I really felt Craig begin to open up. And the more I heard about the onslaught of abuse he got a few years ago, the more I valued his trust. The conversation ranged over many subjects. I discovered he was once a 'prepper' – someone who prepared for societal collapse with stocks of food stashed around the place. There is an entire community who still play this game, in the United Kingdom and especially in the US. I use the word 'game' advisedly, especially for the United Kingdom. In the US prepping is largely centred around firearms. And here, if everything went to hell in a handcart, tins of sweetcorn and a water filter are not going to protect you from a determined mob.

I had to draw us back to the squirrels. Craig began to explain how he became so involved with this project. 'Look,' he said, 'you can't talk about this work without mentioning Esme Kirby. She was the person who got me involved – without her there would still be no red squirrels on Anglesey.'

I had to admit that I did not know who she was. It turns out she was quite a force of nature. Born in 1910, Esme moved with her parents to North Wales before marrying Thomas Firbank in 1935. He became famous as the author of *I Bought a Mountain*, the autobiographical tale of his purchase of a sheep farm near Capel Curig and his process of learning the trade in stunningly beautiful surroundings.

After they divorced, Esme was left with the farm and a growing fascination with the idea of conserving the beauty of Snowdonia. This led her to set up the Snowdonia Society, which gave her the capacity to critique planned despoliation by developers, and also chastise conservation bodies when she felt they were getting stuck in 'airy-fairy ideas.'

'Grey squirrels first made it to Anglesey in the late 1960s,' Craig continued. 'At the time it was noticed that as they arrived, the reds disappeared. We did not know the cause until the late 1990s. It was assumed to be a matter of competition, but then the reality of squirrelpox virus became apparent. It is a horrible disease, kills pretty much all the reds who get it – but infection leaves the greys, who carry it, unscathed.

'By 1998 there were fewer than 40 reds left on the island and Esme was determined to do something about it. And this is where I got involved. She had an attitude of "just go for it", which was great for this, but it did also mean many of her projects had a committee of just one person!'

As Craig spoke, I realised that over the years I have met a number of people with that sort of nature in the conservation world. Esme died in 1999, but not before entrusting Craig to make Anglesey a red-squirrel island.

'It is not just about killing grey squirrels,' Craig said. 'There is no point doing that if there is not a will among the people on the island to support the reds, to manage the land with reds in mind. There is no way I would be given access to the majority of the 720 square kilometres of Anglesey and Holy Island.'

Silently, we both contemplated the impossibility of Craig forming a one-man red squirrel taskforce over such a massive and diversely managed area.

'Do you want to see any red squirrels?' Craig asked.

We had to drive back to the island, of course. Now, I am used to having people wanting to come and see my tame robin or, when I was out in the field, to come looking for hedgehogs. But having an interested guest along was the absolute guarantee of a no-show, so I did moderate my expectations as we headed back to the bridge over the churning Menai Straits. We did not drive for long on the island; Craig was aiming for another supermarket car park.

As I pulled up beside him, he was out and ready. Alongside the shop was a path that led into 'The Dingle'. There are dingles all over the place here – a wonderful word for a wooded valley – and this one soon left the retail zone as trees swept up to the right and the sound of a stream rose from the left.

Craig was very relaxed as to the possibility of seeing a red squirrel. And within a few minutes we saw a couple of people stopped beside the path, staring into the trees. Joining them, it was easy to see why. There, 20m away, on an old oak was a small feeder and on that was an impossibly delicate animal.

I remember the last time I saw a red squirrel; it was on a bird feeder in a suburban garden in Berlin, not far from my brother's house. Again, the relative delicacy of this animal compared to its American cousin was startling.

I could sense the pride in Craig as he watched this animal nibbling nuts. It would not be there without him. He had brought a bunch to this very dingle for release. This did not mean just letting them go, there is a careful process as they are 'soft' released; allowed to acclimatise in large enclosures that were laboriously carried down the dell.

And now, some 18 years since their ancestors were returned to this land, here was a beautiful example, feeding. The nuts were in a box with a hinged lid, and the squirrel had to use its nose to lift the lid, before leaning in and grabbing the bounty. Then the squirrel would face out, alert for danger, and use its dextrous fingers to hold the nut for nibbling.

Now, I had brought with me my longest lens, but still, I have no pretensions to being a true wildlife photographer. My usual subjects are lit by candles and standing in the stalls of ancient chapels – being a choir photographer is perhaps even more niche than being a hedgehog expert! So I wanted to get closer to the squirrel – and this meant we had to play a high-stakes game of Grandma's Footsteps. Every time the squirrel had its head down in the nut box, I could risk a few steps up the hill, aiming to get behind another tree. Over the course of about 10 minutes I got what I needed and the moment I was careless, the tufty ears twitched and it was gone.

As I returned to the path I was full of joy at this sighting, but also troubled. I have laid out my prejudice – I hate killing. And even if I do not hold the Kantian view that the real damage when dealing out cruelty to animals is the harm it does to the soul of the perpetrator, I do wonder what it does to a person.

It was with more seriousness that we continued our walk. First of all, how do you kill a squirrel? And how many?

'In total,' Craig answered, undefensively, 'probably around 7,000. And how, well, I am not a cruel person, I could never leave a snap trap to do its best, and spreading poison in this place would be criminal. I use live traps that I check regularly – no animal is left overnight.

'Our method of killing is simple,' he continued. 'I put a bag over the end of the live trap, open it up, the squirrel runs into the bag, I scruff the bag and hit the squirrel on the head with my priest [a baton used for killing fish or game]. The blow kills, it is over in seconds. The only time I have hesitated was when there was a youngster that did not react like an adult. Just sat there, waiting. It was hard to kill that one.'

It sounds brutal, if efficient. And I bring up the question that has been bothering me, about the impact that this has on the person. Does Craig find that, after all the killing, he has become desensitised?

'Are you asking whether I have turned into a psychopath, or whether I was one to begin with?' he answered, regaining something of his earlier defensiveness. 'I am not into killing, I don't want to kill grey squirrels, I don't enjoy it, but I wanted to see red squirrels back on the island and there is no other way.'

Thus he articulated the problem at the heart of this book – do we sit back and let our ancestors' mistakes rob us of these tufty-eared beauties? Or do we 'defend' islands, killing greys that arrive?

The last grey squirrel killed as part of the control programme was back in 2013. In 2015 four were found. Occasional incursions occur and Craig still reacts to reports of sightings. It seems that grey squirrels are pretty ingenious when it comes to reaching the island. Some have snuck into camper vans or lorries. Some have crossed the bridge, and amazingly some have even swum the channel.

When Craig told me about the swimming squirrel I was rather sceptical. I had watched the water the previous evening – it was a tumultuous stretch, 500m wide at the bridges, with fast currents, swirling eddies and whirlpools. Boats have been lost along there; it is dangerous and there is no way I would swim in it, nor imagine a squirrel making a crossing.

However, the water is not always like that – between the rush in and out of the tide, the water is calm and still, and it was at this time a squirrel was seen making a brave attempt at getting to the island. There are plenty of greys on the mainland, so there is always a risk of another incursion.

I wondered whether he was aiming to eradicate more squirrels from the mainland, but was quickly reminded that killing grey squirrels is not what this is about. Having living red squirrels is the goal. Which meant more than just killing the greys. The reintroduction was an even bigger enterprise – at each site large enclosures were built to house the new reds, to get them acclimatised to their new home. This meant lugging gear over many miles, and repeating it in different patches. Anglesey's woodland is quite fragmented, so to accelerate the reds getting a claw-hold, releases took place in eight different locations.

Nest boxes were also provided, complete with bedding – it is no wonder Craig was so much fitter than me. And of course, there needed to be squirrels. Captive stock was collected from around the United Kingdom. I was interested to find out how much care is put into the replenishment of this species; the Welsh Mountain Zoo co-ordinates a stud book, making sure that the reds do not end up with family trees that are too linear.

The provision of reds is the relatively easy part of this campaign. The work Craig put in with his priest is currently the best way forward for tackling the removal of greys. But there has been growing interest in trying to tackle greys without the application of force.

Contraception is often floated as the solution to the problem of an animal being in the wrong place at the wrong time. Contraception has been a bit like nuclear fusion: one of those ideas that would be amazing if it worked, but practical obstacles meant it was going to take more research.

Contraception for conservation has been used within captive settings for a long time. While zoos, which claim to be an ark for threatened species, need to keep those species alive and breeding, what to do if there are too many? Hormonal control of these populations can often provide a solution.

Though not always – take, for example, the ridiculously cute cotton-top tamarin. This critically endangered squirrel-sized primate has an uncanny resemblance to Albert Einstein. It also breeds really well in captivity. In their natural forest habitat in Colombia, the social groups are very close-knit. Young are reared by the community, and are also the glue that holds the community together.

In captivity, well, what do you do? You can't let them go on breeding as there is not enough space. There are already far more of them in captivity around the world than in the wild. And you can't use contraception because this will cause a breakdown in their society, as the young are the glue. So, you are left having to replicate reality – controlling numbers through 'predation'. Healthy tamarins have to be euthanised.

I once interviewed the education officer for a small zoo, since closed down, and he presented me with a terrifyingly complex ethical question. Their zoo had too many cotton-tops. And while it was a lovely idea to release them back into the wild – which is the aim of many zoos – there is not enough wild left into which they can be released. The next logical step is to pass on the excess to other zoos, but every zoo with this animal suffers from the same problem. Which means they were forced to kill some, just to keep the population stable.

And then came the zinger. If they are going to be killed, he said, why not use that killing to make money, which could then be ploughed back into trying to protect and expand tamarin habitat in Colombia. You see, in the 1970s, tens of thousands of these monkeys were taken from their homes and used in biomedical research, particularly for studies into colon cancer in the US. That trade was stopped, but they are still being used, from captive stock. So why not sell the excess from the zoo into research? The outcome for the animals will be the same. They will die. But their deaths have the potential to allow future generations to be released back into the wild as habitat is reclaimed thanks to the money earned from this trade.

My reaction to this was one of revulsion.

But … if killing for conservation is something that can be agreed upon, is the sacrifice of a few tamarins, for the sake of the wider population, one worth taking? We are back to Trolleyology – would I kill the fat man? Would I actively cause the death of a few tamarins, or passively accept the status quo?

Sometimes, however, it seems that contraception can work. In 2022, a trial from the Animal and Plant Health Agency (APHA) in the United Kingdom has now published results that show, in theory, oral contraceptives can be fed to grey squirrels, and not red squirrels. The last thing we need is for reds to have reduced reproductive capacity at the same time!

Sarah Beatham is the field ecologist who was tasked to find a solution and she has been brilliantly inventive. First of all, there is the hopper – the device that contains the bait. She used one that was 'smart' – it collected data on the number of visits and the amount of bait taken, along with photos and videos of what was happening. Initially the trial took place in a part of the country with no reds, to prove the concept. The hoppers had a 70g door that had to be pushed open by the animal, in order to access the bait. Every time the door was lifted, the date and time was recorded automatically and this was then linked to the evidence from the remote camera that filmed the action. They got images of thousands of grey squirrels getting into the hopper, and once, a mouse ... though that must have been an awfully mighty mouse (or a faulty door).

The next step was to have heavier doors, to see if she could prevent reds from entering. Testing both 200g and 240g doors, she set up trials in woodland areas of Northumberland where there were just red squirrels. During two weeks there were 16,000 photos and videos taken of animals around the hoppers. When they were open the reds fed happily, but when they were weighted, Sarah described the reds as 'visibly frustrated, swishing their tails in annoyance'!

The next step will be to roll out the contraception. Is the plan to eradicate all 2.7 million grey squirrels from the country? Well, no, that would be both practically and socially impossible. There might be ill-will directed at greys from the lovers of reds, and those invested in forestry, but they are also one of the only wild mammals many town and city dwellers get to see. And back in St James's Park, it was not just parakeets that were exciting the crowds of people, it was the grey squirrels too – they have become very acclimatised to humans bearing nuts!

So this story is bigger than the island, then. The body that is looking at the wider picture is UK Squirrel Accord – a partnership of 43 conservation, forestry, land management organisations, government agencies and companies. They have two aims: secure and expand UK red squirrel populations beyond current thresholds, and to ensure UK woodlands flourish; and deliver multiple benefits for future generation of wildlife and people.

The squirrel conflict is a fascinating introduction to the complexities of wildlife management. Well-meaning voices may call for dramatic action, but do not necessarily help the cause. For example the language used by Rupert Mitford, the 6th Baron Redesdale, when talking to the *Guardian* back in 2008, presents at the least, a lack of tact. He described how he had developed a killing strategy for grey squirrels: He described how he had developed a strategy for maximising the kill of grey squirrels, clearing a woodland through shooting, waiting for neighbouring squirrels to move into the recently vacated habitat and hitting them again.

The entire affair was presented with a sort of 'Boys Own' glee, and he certainly did no favours to the more mature conservation plans of others by pointing out that they only used the name Red Squirrel Protection Partnership because there would be much less public support if he declared they were the Grey Squirrel Annihilation League. He was brazen enough to conclude that they actually do nothing directly to help the reds, other than killing the greys.

The work of Craig, however, is far more considered, and while I still feel uncomfortable with all of the killing, it is being done with a goal in mind ... and the goal is not the accumulated bodies of the target, but the flourishing of the conserved, something that all conservation projects must keep in mind.

Deer

Wytham Woods is one of the most amazing scientific playgrounds in the world. Just beyond the Oxford ring road, this 430-hectare woodland has seen the development of scientific ornithology, has the densest concentration of badgers in Europe and has been the scene of at least one murder mystery.

For that last contribution we have to thank *Inspector Morse*. The timing of the episode appearing on television was poor, as it happened just before I was about to embark on some fieldwork in and around the woods. Being called up by one of the leading mammal scientists in the country, David MacDonald at the University of Oxford, and asked whether I would like to undertake a few weeks of small mammal trapping early in 1989, was too good to miss, especially as he carefully said that this would coincide with the interview for the DPhil position I was applying for. While it would make no difference to the outcome whether I did the work or not, at least I would be in the right city.

There were three shifts that needed to be completed each day, he explained. Each shift required me to inspect the live traps that were gridded through the field margins – we were looking at which management regimes best helped small mammals. So I got to spend three sessions a day meeting many mammals – mice and voles being the most common. I was told that while the traps on the shifts that took place overnight – from 6.00 p.m. to 2.00 a.m. and then 2.00 a.m. to 10.00 a.m. – would be pretty full, I should be able to get some sleep on the research station's sofa after the daytime shift, as there would be very few animals to slow me down.

It was hard, but fascinating, and spiced with the potential threat of homicide that increased its presence as tiredness grew. Oh, and I didn't get the DPhil job … what a different life I would have led had that happened, doing research in an academic setting.

But the woods – amazing as they are and full of well-studied life – are not in any way wild. They are carefully managed and a large part of that management is the control of the deer. To preserve the integrity of the woodland there is a deer-proof fence around it. But there are also deer within, and each year they have to be culled, the fallow deer in particular, to prevent the damage that overcrowding would cause to the flora.

I thought it would be interesting to spend some time with the woodland managers, but this was swiftly denied. There is a great deal of sensitivity around the killing of deer for conservation. It is not just the plant life that suffers from deer – research has shown how the loss of low vegetation, such as bramble, removes nesting sites for many small birds and their numbers were dropping.

Deer are reckoned to be as great a threat to woodland ecology as climate change. I find it fascinating that shifting baselines are at work in this setting too. Deer numbers have been slowly growing, and in so doing, equally slowly increasing the amount of damage they do to woodland such that we don't notice the change as the landscape is degraded.

It is not just in enclosures such as Wytham that deer are a problem. One of the great leaps of ecological imagination has been the superficially simple idea of allowing nature to just get on with its own business – to rewild. This simple idea has proved to be divisive, as different people put different spins on the process, but at its heart it is something to be explored and encouraged.

The reality is that rewilding needs some management as there are no complete ecosystems left – by which I mean there are no predators left who can maintain anything approximating a natural balance. Which means people need to kill deer, who otherwise prevent trees from growing and prevent the rewilding process from taking place.

Many of the people who are instinctively attracted to the idea of a self-willed landscape are also instinctively revolted by the killing that needs to take place to allow this to happen. An advocate of rewilding, George Monbiot explored the uncomfortable nature of this on a television documentary he made about veganism, in which he joined deer stalkers who help the forests to grow by killing deer. With careful supervision George takes the shot that kills the deer and is clearly sad, though the look of delight as he munches on some venison after it is butchered cannot be denied.

George was late to the vegan world – when I first met him it was me that was vegan and he the avid carnivore. I remember making the point to him about the impossibility of reconciling a real love of the environment with a meaty diet, but failing to win him over. So it was with delight that I interviewed him on the stage I run at a festival in Oxfordshire – and put this to him – and his response was 'I admit it; you were right, I was wrong.' I turned to the packed tent and asked if anyone had filmed that admission – as it would be something to treasure – and saw my wife at the front, filming. It was only later when we watched the footage that it revealed all she had caught was, 'Did anyone film that?' But there were plenty of witnesses if you need affidavits!

CHAPTER 4

Lundy

Trying to find a consistent answer to the question of how many islands there are on the planet is tricky. This is due, not least, to definitions, some countries defining an island as simply being able to support life of some kind, while others don't count islets and cays. But if we are to take the World Population Review website at its word, and it does seem to be one of the more reliable sites, there are around 670,000 islands on our planet, of which 11,000 are permanently occupied, providing homes for 730 million people.

The two countries with the most recorded islands are Norway and Sweden, both having more than 200,000. If Canada had different criteria and counted in the same way as these two, it might just top the list.

Islands are particularly vulnerable to the arrival of new species of animal and plant. And it is therefore on islands that many of the decisions about whether we should be killing for conservation have to take place.

To get an idea of the scale of the problem, I suggest you have a look for the Database of Island Invasive Species Eradications (DIISE). This online resource gives us a terrifying insight into the number of islands that have recorded problems with invasive species. But it also comes equipped with a heavy dose of hope, saying, 'Protection of threatened biodiversity by removing invasive vertebrates from islands is becoming a powerful and widely used conservation tool.'

The research paints a pretty grim picture – in the last 500 years, invasive predators have caused the extinctions of 87 bird, 45 mammal, and 10 reptile species on islands. That is over half of all extinctions in these groups. A further 596 species are at risk of extinction because of invasive mammals – cats, rodents, dogs and pigs being the most serious threats.

It is no surprise that the most vulnerable populations are to be found on islands. And as many species on islands have high evolutionary distinctiveness, this means that invasive mammalian predators are an important driver of the loss of global phylogenetic diversity (diversity as a result of distinct ecological relationships).

Evolutionary distinctiveness is an interesting concept. The EDGE of Existence programme, based at the Zoological Society of London, has been important in raising awareness – Evolutionarily Distinct and Globally Endangered is where EDGE comes from. Now many of the species highlighted by EDGE are a little odd, and very often overlooked. In the context of the tree of life, many are out on branches of their own, and as such they contain the genetic history of everything that has led up to that one species. So the loss of one of these species represents a bigger hit to the total sum of genetic diversity. Some might dismiss these odd-bods of the animal world as 'evolutionary dead ends' but that is to miss what they represent.

So islands are simultaneously uniquely important, and particularly vulnerable to the additions of new species. This is why action is deemed so necessary.

In the process of ensuring information is shared and the best approaches are used, the DIISE has recorded more than 1,200 cases of attempts to eradicate invasive species, and in the 1,000 or so cases where results have been recorded, the success rate is around 87 per cent. The closest of those I could see was Lundy.

Lundy, off the north Devon coast, has a story to tell – in fact there are many, ranging from pirates to twentieth-century would-be kings. But the story that fascinates me most concerns rats.

Many species of mammal were introduced to Lundy by Martin Harman, who took control of the island in the 1920s. He was certainly a character, who eventually became too much of a character when he started minting his own currency. Sika deer, Soay sheep and red-necked wallabies all made the island their home thanks to Harman's generosity.

And while there are issues with the deer and sheep, and the many other mammals that have been introduced (the wallabies, sadly, are no more), it is the rats that have caused the most consternation. He did not need to introduce rats; they had made their way over with earlier visitors, though there is no record of who was responsible.

The natural inhabitants of the island are birds, most famously those avian clowns, the delightfully charismatic puffins with their multicoloured beaks, enthusiastic flapping and almost penguin-like walking.

This was going to be an adventure, and adventures require a companion where possible, so I suggested to Miriam Darlington that she join me. Miriam has written beautiful books about otters and owls, and also has a monthly column in the *Times* newspaper about nature … and this was the perfect opportunity to explore an area beyond her usual patch of south Devon.

So we hatched a plot, and I got in touch with Rosie Ellis, the warden on the island. Finding a time that worked for us all narrowed it down to a Saturday in July 2022. But it was not until the night before that I did some fundamental research, such as how long the ferry would take … I had been putting that off for the very good reason that ignorance is a form of bliss.

I love being on, and in, the sea. I love being on boats … for about 25 minutes. For that long I can cope with anything, but by the 26th minute my body starts revolting. Years back I had spent two full days out with the Cetacean Research and Rescue Unit, up on the north-east coast of Scotland, looking for, well, cetaceans. My fears back then had been alleviated by the pill that volunteer Izzy had given me, with the assurance that this was 'good stuff'. It was! I had an amazing time.

But for the two hours to Lundy, I had no drugs, just the determination to not embarrass myself. I should not have worried. The weather was dramatic, in that it was clear, smooth and beautiful, for both legs of the journey. Nevertheless, I did find my spot as close to the front of the boat as I could get and stared fixedly at the horizon for nearly the entire time, watching the land sigh into the sea.

Miriam is good company – lots to talk about and also very happy to let a companionable silence sit comfortably. Every now and then one of us would lift binoculars and exclaim, as we spotted razorbill, guillemot, Manx shearwater and gannet. The others are great but the drama of a plunging gannet is hard to beat. And we were both waiting, knowing that there were dolphins around. I know I should not just concentrate on the obvious species … all of nature is wonderful … but dolphins … they are special.

Our ferry, the *Oldenburg*, pulled in at the southern tip of the island. Launched in Germany in 1958, this lovely old boat has been ploughing the Bristol Channel since 1985, and I did not feel ill, and for that I am forever grateful. Rosie met us but was distracted – she had been hauling ropes to secure the boat, and was now in the role of first-aid officer, as someone had fainted on the crossing. So she asked us to make our way up to the inn, grab a drink and wait for her. That did not seem too onerous. Though as she bounced past us in the Land Rover, and as we paused, again, on the path up the cliff, I did see the advantage of getting a lift.

As we began the ascent I noticed something almost hidden, and rather out of place: a rodent bait box. I took a photo; this was why I was here. But I did not know it would be so obvious.

The walk up was longer and hotter than I had expected so the cold drink at the top was well appreciated. As was the view - we could see both Wales and Devon from where we were, really making our place in the Bristol Channel apparent. The waters around the island are remarkable. The

9m tidal range, coupled with the great mixing of estuarine water and the Atlantic, churns up nutrients galore – resulting in the area around Lundy becoming the United Kingdom's first Marine Nature Reserve, first Marine Conservation Zone, and being recognised internationally as a Marine Protected Area. The island is also recognised as a Special Area of Conservation due to the rich nature of the reef systems with which this granite nodule is associated. It is also a 'no-take zone' – no fishing allowed.

All of this makes it a wonderful spot for birds. Puffins come to breed and feed on the sand eels, many of which end up posing in the beaks of parent birds, ready to stuff into the pufflings … and yes, that is the real name for a baby puffin. While the adult puffins usually manage to carry around 10 fish at a time, the record is set at 126 – that is some beak. Islands like these are also a magnet in spring and autumn for migrating birds, and the twitchers who flock to see them.

Rosie has an amazing job. While she trained as a marine biologist, she has been employed by the Landmark Trust for three years so far to oversee the island – conservation, tourism, public relations, track maintenance, boat-rope hauling – she does it all. As she walked us to a the small buggy, like a rugged golf cart, with which she was going to give us a whistle-stop tour of Lundy, she pointed out that while many people apply to work here, you do need to be of the right mindset – it is not easy. And while the helicopter does take only 6 minutes (I have to say, even though the boat had been smooth … that did seem like a more attractive option) you are definitely offshore.

'Let's head to Jenny's Cove,' she said as we clambered into the vehicle. About halfway up the island, Jenny's Cove is one of the famous spots that visitors aim for; it is where puffins and guillemots nest. So I was excited, because while I was here to ask serious questions about killing for conservation, seeing a mass of these auks is always a thrill. Seeing my excitement, Rosie was quick to dampen it with a reality check, saying, 'But you should have been here a week ago. It is the same, every year, just as the schools get ready to break up and we are flooded with children who would love to see a puffin, they head out to sea; breeding finished, job done.'

The buggy kicked up a cloud of dust as we bounced north up to the halfway wall – which you get to after the quarter wall and before, you guessed it, the three-quarter wall. But other names are less prosaic, and Jenny's Cove is bounded by The Pyramid and the Devil's Chimney.

One of Rosie's volunteers was there with a telescope pointing towards the cliffs, which I could already see as a site of activity; birds careering in and out of holes in the steep sides of the cove.

'Every year we have people wanting to get close to the birds,' Rosie said. 'But it is not great for the birds to be disturbed, and really, you turn up with a phone and want a photograph of a puffin, you are better off buying a postcard. The seabirds are protected by law – we are not a zoo. The wildlife is free to come and go and hide as they please. And most people who come to the island know that and appreciate it, so we don't get trouble. Just a queue of people wanting to look through the telescope.'

My binoculars did a fine-enough job, but I swiftly gave up trying to get a photograph. I could see the puffins frantically flapping and the guillemots, well, Rosie was quite able to turn me on to the wonder of these mini-penguin lookalikes. They are remarkable. While many birds when they dive for food at sea use the momentum of their flight to carry them down, these little auks swim down and down typically to 30–60m but have been recorded at 180m. Add to that fact that the males are the ones who teach the young the ropes of life as a guillemot, and you have something even more inspiring than the clown-faced puffins.

Puffins do have a claim to the island equalled by no other bird, as Lundy is derived from the Scandinavian for puffin – Lund. To add another layer of intrigue, the Latin name for the most important bird on the island is *Puffinus puffinus* – the Manx shearwater. I know, it makes little sense. The shearwaters are like mini albatrosses, they are 'tube-nose' birds – the puffins are auks. And just to confuse us further, the puffin actually got its name because it nests in the same style as the shearwaters, which were known as Manx puffins until the seventeenth century.

Now that we'd had our fill of the charismatic birds, it was time to find out more about the rats. Lundy's rats were an interesting bunch. The past tense is relevant as the island has been rat-free since 2006. When they first arrived on the island is not known, but what researchers did find is that Lundy was a rare example of a habitat where both the brown and the black rat were living together. Rats get to islands inadvertently; in cargo or from shipwrecks. I doubt that anyone has looked at a new home and thought, gosh, what we really need are some rats. They do not have the sentimental attraction of a hedgehog for UK ex-pats in New Zealand, for example.

Black rats are fascinating. Yes, they have got a bad press with their alleged involvement in the spread of the Black Death and their brilliant ability to infest places. But they are gorgeous too – staggeringly intelligent, excellent at climbing. I remember being introduced to a tame black rat that was kept by a manufacturer of poisons as a rather unfortunate poster-beast in adverts. It was massively sociable with people and happily climbed up my arm, whiskers twitching, inquisitive eyes twinkling.

Black rats also happen to be one of the rarest mammals in Britain. They are smaller than the more familiar brown rats, who will also eat them. They have been caught up in general control of rodents around the country and it is possible that

Lundy was the last viable population of this animal in Britain. But a decision was taken to kill them.

I am not saying that was the wrong decision, though there are some who did actively protest against the killing. In December 2002 the campaigns officer at Animal Aid said, 'This is an attempt to restore ecological harmony through wholesale slaughter. It is not the way forward – in this case more controls on commercial fishing, management of pollution and protection of breeding sites would help boost the seabird population.' Demonstrations were held, but it is a challenge to whip up sympathy for rats, even if you have an absolute opposition to them being killed.

More importantly for me, though, is the need to recognise that an active choice was made to kill these animals; that on Lundy, the rats were deemed less worthy than the birds.

The reason an extermination campaign was launched in 2002 was simple. There were just 13 puffins in 2000, down from 3,500 in 1939, and there had been no successful breeding since 1986. The British Isles is home to over 90 per cent of the world's Manx shearwaters. Their numbers on Lundy were down to 166 pairs in the same survey, and they had not produced any chicks since 1959. There were two main reasons behind this: loss of sand eels – a main food for much of the bird life on the island – and rats, who would eat eggs and chicks.

As is the case in many of these stories, the problems are multifarious, but the options for action are often limited. Getting sand eel numbers back to the sort of level where the birds of these islands could flourish again would require international control of industrial fishing alongside actual action to prevent climate change.

Much easier to remove the immediate threat on the island. Not necessarily easy, just easier.

The Lundy Seabird Recovery Project ran between November 2002 and March 2004. A lesson here in good presentation was that the project was all about the seabirds recovering, not the rats dying. This was one of the PR traps the work on the Uists fell into. The work was a collaboration of the National Trust, who own the island, the Landmark Trust, who manage it, the RSPB, which has the conservation knowledge, and contractors from New Zealand.

'Starting in January 2003 the team set 1,923 bait stations loaded with poison,' Rosie explained. 'They were spread in a 50-m grid across the entire island to maximise the chances of them being encountered by a rat. They used a cereal-based wax block with difenacoum as the active ingredient.'

While killing the rats was a priority, monitoring what was going on was also crucial, so an additional 2,000 stations baited with a piece of soap, candle, or small wooden stick dipped in vegetable oil were set up – so that if rats were present the teams could tell from the teeth marks they would leave as they gnawed on the bait.

And that was just the first phase. As some rats were still found, phase two was started in December 2003, running until March 2004. A different formulation of poison was used as there had been some concerns about the durability of the initial choice. Throughout these sessions of work, all the stations were checked every two to three days, poison refreshed where necessary, and and corpses removed for safe disposal to keep them out of the food chain.

That was not the end of it, of course. Two years of thorough checking followed, and by 2006 the island was declared rat-free. And then? By 2023 the ornithologists counted more than 10,000 pairs of Manx shearwater and more than 1,200 individual puffins. Other species have benefitted too – pygmy shrews, those hectic snufflers, have increased in number, as have one of the smartest birds, the wheatear. The male wheatear in particular looks so dapper, and flaunts the feature after which it is named … no, not a straw-coloured marking on the side of the head, but a white rump … this gorgeous bird's name is derived from 'white-arse'.

Were there any negative impacts of removing the rats? It was argued that a surge in rabbit numbers following the extermination was the result of rats being removed, but by the time I got to the island, the rabbits, having been hit by another wave of myxomatosis while they were already under the hammer of haemorrhagic virus, were almost all gone.

'We are always concerned about the impact poison might have in the environment, so the rodenticide is always placed inside bait boxes,' Rosie said. 'Now these stop most other animals from getting in, but you will sometimes find the teeth marks of the pygmy shrews. When we had the 2,000 bait stations on the island, they were at 50-m intervals. As the shrews have very small home ranges, this means that while a few individuals may get a small dose, there would be plenty of shrews not near a bait station.'

As for the rats, the work does not stop – it never will. Rosie was serious as we headed back south. When we got to the buildings in the village she took us to a shed that contained stacks of the bait stations that I had seen down by the pier. She pulled out a bar of something that looked a bit like soap and asked me to smell it. I got a very distinct whiff of chocolate; apparently rats cannot resist the heady wax and chocolate combination. But no poison, I asked?

'We don't want to leave poison around the island,' Rosie said. 'The risks are very small, but I would not want to take them. And anyway, we don't need poison, we just need these.'

So the bait station I saw, and the other 59 around the island, are checked every month, and have been since 2006, for evidence of rat tooth marks. 'So far we have not had a single positive sighting, but we cannot afford to become complacent. And we don't rely just on these, we also ask that all visitors report sightings, though so far, every single record has turned out to be a pygmy shrew, rabbit or water rail.'

There is a good relationship with the ferry company that brings out the streams of visitors from Ilfracombe and

Bideford, as this would be the easiest way for a rat to get to the island. The ships are checked, carefully. Other visitors cannot moor their boats to the pier, but can moor in the bay, again, reducing chances. On top of that Rosie works with the boat clubs on the mainland, ensuring the concerns of the island are understood.

The one variable that cannot be controlled for is a shipwreck – though modern navigation systems have reduced the occurrence of collision. In fact it is amazing that there were not more collisions with the island in the past, considering it is often covered in fog, and is in the middle of the routes to South Wales and Bristol.

'We have had one sighting in the three years I have been here that did generate real worry, and we went to the first level of action,' Rosie continued. 'We got the team back out to place poison in these bait boxes across the island. We found no evidence, but as I can't stress enough, we cannot take our eyes off this, ever. Dreams of rats wake me in a cold sweat at night. The only thing that will matter when I am dead and gone is that Lundy is rat-free.'

That last statement hit home, it was said without hyperbole, just a fact, and true measure of the commitment of this person.

I asked Rosie about the rabbits, because the rabbits are an 'alien' species too. 'There are records of rabbits being taken from the island for fur and meat as early as 1183,' she said, as birds swirled in the bay. 'Though the story I like best is how they were deliberately imported at the time of the building of the castle in around 1250, which in turn is related to a long history of piracy and angry kings having people ripped limb from limb by horses. The rabbits were deliberately bred in a warren to help pay for the construction.'

I was building up to my 'gotcha' moment. The rabbits are an alien mammal, just like the rats. Yet why were the rabbits given a pass? Why was the seemingly arbitrary decision made to keep one alien alive, while killing the other? Was it just because of the puffins? (Now that might seem like a bit of a leap, but one of the reasons that puffins do so well on Lundy is that they utilise the rabbit holes as nesting burrows.)

'How do you think puffins made burrows before the rabbits were here?' Rosie responded. 'They use their bills to cut into the soil and their feet to shovel it out, a bit like a dog. And yes, they will use the burrows of rabbits, and even Manx shearwaters, but they can manage well enough on their own.'

Rosie and her team have to make so many choices – which animals are allowed to stay being a large part of it. House sparrows, for example, were introduced, and it was amazing to see healthy flocks around the farm and inn. In fact, it was a rather painful reminder of what we have lost. It is so easy to forget, when we see the sparrows chattering away in the privet hedge and think what a lovely flock. But what I have been seeing is trivial – a sorry relict of what there should be on the mainland, hopping agitatedly between our gardens.

I asked her how this decision is made, about where the line is drawn. A familiar question in this book, but one she gave a very straightforward answer to. 'The island has been dipped in aspic, setting it at around 1970.' I was surprised. 'The Landmark Trust have to draw their line – they have to choose at which point in the island's history they want to maintain buildings. And they chose 1970 – there had been little "improvement" to the buildings up to that point. They are now cosy, and rodent-free, oh, the buildings were where the infestations used to be really quite dramatic.'

They have to manage the Soay sheep, the goats and the Sika deer – all imports from a less enlightened, but nevertheless enthusiastic, time. The aim is to not turn the clock back to some imagined Edenic paradise, but to try and maximise the benefits for, mainly, the birds that make the island so special – and in fact gifted the island its very name.

Rosie had been massively generous with her time and we had to say goodbye. As we were about to head back down to the boat, Rosie suddenly remembered something. 'You really ought to track down Jaclyn Pearson at the Isles of Scilly Wildlife Trust – she is doing some fascinating work down there. We got her to visit Lundy so we could learn from her.'

Miriam, who had been generously in the background, aware that she was piggybacking on my research trip, stepped forward at that moment. 'We have to go,' she said. 'I used to live on Scilly – come on – it would be a great adventure.'

We turned towards the small shop and the possibility of a very necessary ice cream; Miriam was beaming. Not only had she plenty of material for one of her magical *Times* nature diaries, but she had also fallen a little in love with the island and was plotting a longer visit next year.

I had forgotten that a two-hour ferry ride out to the island would tend to indicate a similar length of return journey, and was suddenly very aware of having eaten an ice cream as we boarded the *Oldenburg* again. We managed to grab similar seats to last time, near the front. We were sunned, but not too much, tired, but not too much, and let the gentle movement lull us as the light changed. I was still slightly on twitch for a dolphin, but had low expectations until, nearing Ilfracombe, Miriam shouted – she saw one first – then as I looked up, another, and another, and we were, for a blissful five minutes or so, in a pod of dancing mammals.

The birds had done a good job of luring me over to their side, but leaping dolphins? Come on – what can compete with the utter joy that they generate!

A few months after my visit to Lundy, the importance of Rosie's readiness was made apparent when another UK island went into emergency action. In December 2022, a fishing boat hit a rock and ran into the island of Skomer, off the Pembrokeshire coast. After the crew were safely rescued and the area surveyed for oil spillage – fortunately none

– the next step was for Lisa Morgan, from the Wildlife Trust of South and West Wales, to run through the measures outlined in their biosecurity plan. This 70-page document was produced to provide an extra layer of security for Skomer, and the two smaller neighbouring islands of Skokholm and Middleholm. Most of the plan is to do with everyday biosecurity – such as making sure that visitors check bags before getting onto boats out to the islands, and the monthly checks of non-toxic surveillance baits. Rats must be kept off the islands.

As with Lundy, though, there is a plan of action for moments like this. These islands are holdouts for puffins and Manx shearwaters, and the numbers on Skomer alone are just astounding, with 43,000 puffins and 350,000 breeding pairs of shearwaters.

A shipwreck is the most likely way that a damaging predator will make it onto the islands, so this warranted a full response. Lisa's diary of the next few days, published on her Wildlife Trust's website, is a fascinating read – a valuable insight into the need for a rapid reaction. Within 48 hours she and her team had got bait stations set up and some of them set with poison. They could not set them all immediately as the winter days are short, and working in the dark would be dangerous. The stations are simply 75cm-long plastic pipes on wire legs, which hold cereal-based poison bait blocks securely and keep them out of reach of birds and rabbits.

By day three all the stations were set. The plan states that stations were needed every 50m (so 4 per hectare) across an area extending in a 1,000m circle with the location of the boat in the middle. Obviously a chunk of that is underwater so not needed, but this is the density at which they need to be placed. The entire grid of 63 poison bait stations, along with 10 surveillance stations (irresistible chocolate-flavour wax again) and two trail cameras needed to be checked twice a day for five days, each circuit taking around three hours.

In addition they called in help from the RSPB, which has two specially trained sniffer dogs, one of whom was on hand to work this island. So the spaniel 'Jinx' was employed on day four, to quarter the island. By day seven the absence of tooth marks in the bait stations and the lack of reaction from Jinx gave Lisa and her team the all-clear.

She concluded her diary with, 'This is the first time we have ever had to put our biosecurity plan into action and I hope we don't have to do it again anytime soon. It was hard, cold work and incredibly difficult on Skomer's fragile terrain, with limited daylight and only two staff available. But it was also a very valuable learning experience, with lots of problem solving along the way, all leading to a better, fine-tuned plan for the future, learning what will assist seabird sites like ours across the UK, all of whom would have to do what we did if faced with a similar situation.'

Dormice

While I *know* that hedgehogs are the cutest of our mammals, there is strong competition from our dormice too. The hazel dormouse is native to these shores and endangered, relying largely on coppice hazel woodland for survival.

But they are not the only dormice to make their sleepy home in Britain. *Glis glis*, the edible dormouse, also known as the European fat dormouse (I am not sure which is better to be honest, I think it should petition to simply retain the scientific binomial) was introduced to Britain in 1902 by Walter Rothschild. Tring, in Hertfordshire, is well outside the usual range of this species, used to warmer climes, but it has held on and slowly spread through the Chilterns and beyond.

Clearly we have learnt nothing about releasing animals into this country with the aim of 'improving' things – though this dormouse has yet to receive the levels of hostility as the grey squirrel. However, given time, it might just end up back in the pot ... there is, in fact, a pot specifically made for the edible dormouse; called a glirarium, this terracotta prison was commonly used by the Etruscans and Romans.

We have not quite got there yet, but they do cause trouble. They are bigger than our native dormice – they can weigh up to 300g (10 times the weight of the hazel variety) and are half the size of a grey squirrel. Unlike our hazel dormice, these bigger beasties will make their homes in our homes, helping themselves to our food and paying little attention to the toiletry conventions. Scampering around at night, setting off alarms, gnawing on cables – they are losing friends.

Their spread is being monitored as they traverse the Chilterns, and set up populations further afield. And while their pest status causes them to be culled, it is that their diet is not entirely nut-based that raises conservation concerns. In fact the researchers who have been following their progress have found that between 10 and 54 per cent of bird nests in monitored boxes were preyed on by edible dormice each year. This, coupled with the economic damage to forestry and the costs of excluding, or removing, them from homes, means that patience is wearing thin.

There is a complication with these animals. As they are listed as an alien species under Schedule 9 of the Wildlife and Countryside Act 1981, it is illegal to release them into the wild. So if you capture one, you have to kill it – despite all dormice having European protection under the Bern Convention.

So either through a misunderstanding of the law, or through understandable squeamishness, people often do not kill any *Glis glis* they capture and instead find places to release them, at a distance so as to discourage a return visit – thereby facilitating the animal's progress around the country.

A conservation cull has yet to kick in, but as the substantial nest predation shows, there may well come a time when more robust action is needed. Or will it be one of the species, like the ring-necked parakeet, that we accept and even enjoy for its injection of diversity?

CHAPTER 5

Scilly

Getting to Scilly was not as easy as I imagined it would be. The need to line up health, weather, accommodation, travel and, most importantly, Jaclyn Pearson, stretched my brain and had me imagining what it would be like to have a PA. In the year I worked for the BBC's Natural History Unit in Bristol I was amazed that when I needed to go somewhere, someone else would sort out logistics and present me with a train ticket and timings. This time (well, every other time really), it was down to me.

My first attempt was floored by a heavy cold. On the second attempt everything was in line, apart from me taking too long and there being no flights available (I was attempting to do this in the winter, failing to take into account that the boat does not run). The third time – at the very start of February 2023, all the pieces came into place. I had assumed that Miriam was joking when she said she wanted to join me on another jaunt, so had only sorted out logistics for me. Luckily she is better at this than me, and within an hour had let me know that she had got the last room in the hotel and a place on the plane.

I am always thrilled with a road trip that takes me to Devon as there is a very special person there, who lives in one of the most gorgeous parts of the country. When I was first told about the village of Staverton, just outside Totnes, I was assured that it was very like the Shire from *Lord of the Rings* – and so it is. The only real failing is the fact that the houses all seem to be above ground. The special person is Silva, and Silva has given me the great honour of being an 'Odd-Father'. My absence of belief in any god meant that godfather was out of the question, so she is one of my odd-daughters and I am her odd-father. And she is amazing. I just wish she lived closer to Oxford. So when in the area I make it my job to pick her up from school and take her home. I love her family too, but the special time with her is important.

After spending a good hour catching up with Silva, and being jumped on by her little brother Skylar – who is a delight of muscle – it was time to head over to Miriam's. She lives on the edge of Totnes, and has a living room that looks out, through the branches of a silver birch, to the hills on the far side of the suburban valley. After eating, she lit the fire and we were joined by her daughter Jenny, who thought it funny to test us old ones on a vintage set of Trivial Pursuit cards. I was amazed at how much nonsense was sitting in my head, just waiting for the prompt of a question. I do worry what that all might be blocking … maybe a hidden talent for the guitar has been subordinated by my apparent knowledge of 1980s pop music.

To get to Scilly meant a two-hour drive to Land's End. As driver, I put on a playlist that trickled along in the background as we chatted – not a single moment of 1980s pop music. I might have failed to win Miriam over to the delights of Olivier Messiaen, but we both agreed on the scatterings of folk music that intruded as we tried to resolve the world's problems and talked about plans for her next book. As soon as she started to describe a journey around the United Kingdom looking into the wildlife of churchyards – in particular the most abandoned ones – I could see it working. Her ability to poetically weave a narrative around natural history truly triggers my jealousy.

The flight to Scilly is very short – just 28 miles south-west of Land's End. As we took off in the tiny 12-seater plane from the end of England, Miriam pointed down to the abandoned tin mines that tell a story of this land, linking right back to the Romans. I was craning my neck to try and see the amazing cliff-face stage of the Minack Theatre. And then, in between the pilots I could see land again; the edge of Scilly as we skirted in above the cliff and rolled to a noisy halt. I had forgotten just how much noise these little planes made for the passengers.

After the bus ride to our hotel in Hugh Town – they do have good taste here – and a brief refresh, we regrouped. The weather was not tropical, but the island demanded attention and with us staying only two nights, this afternoon walk around the coast path of the batteries was necessary. Miriam was a great guide – this was her old stomping ground. It was amazing to see the relics of fortifications, from the Napoleonic War and Second World War. This was an outpost. And while I would never want to be in a conflict, I can think of worse places to be stationed than here, looking out to the Atlantic, with America the next landfall.

The walls of the Napoleonic battery have stood the test of time, and are now becoming one with the cliffs, allowing succulents to grow in the cracks. Aeonium, the tree houseleek, creates rosette brooches on the walls. The more recent metal structures are showing more dramatic signs of weathering, with rust clinging to the surface like lichen, in orange and purple, brown and blue.

Even on a grey day that was darkening rapidly, I could see the appeal. Miriam was just in her element – this is a place with which she is so deeply connected. Five years' teaching and having children on the island have left their mark.

The islands – all 145 of them (depending on when a rock is an island) – are low-lying and largely warm, thanks to the North Atlantic current gifting them heat from the Gulf Stream. This has led to their most famous industry, after tourism – that of cut flowers. As we walked around the battery, Miriam stopped and suggested I take a deep breath of the small double-headed daffodils that were peppering the side of the path. I love scent – not perfumes, at least not most perfumes, they make me feel ill – but the scent of flowers. I am one of those people who will wander around a garden led by my nose as much as my eyes. The wash of fragrance from these narcissi was close to overwhelming

– sweet, not light floral as the shape of the flower might suggest, almost leaning into a hyacinth that has been hanging out with a fragrant jasmine for a while. This was an added reason for our walk taking longer than the mileage would suggest. And a considerable improvement on the last time she told me to smell something – receiving a sample of otter spraint through the post was one of my more unusual experiences.

February is not a busy time for tourists – the hotel was down to a skeleton crew, and was using the quiet to make lots of noise redesigning the dining room. There were only two bars on Scilly open and offering food. The first night we ate at The Mermaid – a sturdy stone building designed to withstand the full force of the weather by the pier. Inside was a clutter of boat things, and a few locals. A pool match was happening later, so our early arrival was fortunate.

After ordering food, I asked Miriam about whether there was a distinctive Scillonian accent – this got us talking with some of the other folk and after it became clear we were only there for two nights, we were told that the pub was shut the next night because the hotel a few doors down had its weekly Queers night. Miriam was delighted – telling me how this was such a lovely transformation, how in her day, more than 20 years ago, there was quite a macho culture, and it was great to see the community opening up to the diversity of people on the island. There was a moment of confused silence from the bar manager before the penny dropped for her. 'Quiz night,' she said, more clearly. … So maybe there is a bit of an accent.

The next morning Miriam headed off on a long walk and I went off to meet Jaclyn. Everyone I had met and told why I was here had said the same thing: 'Oh, Jaclyn is wonderful.' So much so that I did wonder whether this was a bit of a wind-up. No one can have such a clean record, especially on an island.

She arrived a little after me, apologetic, and we headed to a window seat in the hotel bar. We started talking immediately. Coffee came and was cold by the time we drew breath long enough to take a drink. Everyone was right: Jaclyn was an absolute fount of knowledge and good humour. She later explained how important the latter was – luckily for her it comes naturally. 'You know, if I was to bump into someone in the Co-op', she said, 'while tired after a long day, or just generally grumpy, and respond to a question with a grumble, well, that could undermine so much of what I do here. The social side of this sort of wildlife work is underappreciated at your peril.

'You said in your email that you had already been to Lundy,' she said. 'Well, there Rosie and her team have it easy.' Jaclyn was smiling as she said this – clearly there is no 'easy', just different levels of trickiness.

What she meant was that on Lundy the only residents were those working for the Landmark Trust. And while there were visitors, as we have seen, there are very clear biosecurity measures in place.

'So much of this work comes down to understanding people's values,' Jaclyn continued. 'I have worked with wildlife for most of my life, I have worked for the RSPB and the Wildlife Trusts. My values are deeply rooted in that world. But it would be arrogant in the extreme to move to a community and assume everyone felt the same.'

Successful communication and cooperation with local people leads to successful conservation. This is something at which Jaclyn, with her boundless bonhomie, is very good.

'Rats – it is such a shame as they are amazing creatures, but they do cause trouble,' she said. 'Now, there are lots of islands here and not all of them have a problem with rats. The rodents only become an issue if there are breeding birds. People living and visiting the island also adds to the trouble, as they can inadvertently introduce an unwelcome visitor.

'I was an aviculturalist working in Staffordshire, and I came to Scilly on a sabbatical,' she said. 'And then I met Tristan, and he said he would be willing to move back to Staffordshire with me and I just pointed to where we were and said no. I mean, look at this place. I got a job with the local council then some part-time work with the Wildlife Trust and a bit with the RSPB. And discussions started in 2010 about trying to do something really ambitious – removing rats, with a view to allowing storm-petrels and Manx shearwaters to return. It was a new idea, trying to remove the animals from an island with this many people already living there.'

There has been a serious problem on Scilly – the overall seabird population fell by over 30 per cent between 1983 and 2016, when there were fewer than 8,000 breeding pairs. Historically the population of seabirds was in the hundreds of thousands. Nevertheless there was a long lead-in – the idea had to be assessed, the funds raised, but then Jaclyn got a job at the RSPB to make it happen.

'The Isles of Scilly Seabird Recovery Project. You see, even in the title, we were working carefully to make sure that what we did was clearly understood. This was not just a project to kill rats. No rats and still no birds – that is not a success. It might seem blindingly obvious, but there has been a tendency for some of these sorts of adventures to focus on what you are trying to get rid of. Of course, you know about this. Up in the Uists they were measuring success in dead hedgehogs to begin with, weren't they?'

Yes, as I explained earlier, that pivot of presentation, from dead hedgehogs to living birds was the point at which the wider wildlife community was able to be far more supportive of the work. Oh, and the fact that they stopped killing hedgehogs helped too!

'Now, this project was not aimed at ALL of Scilly,' she continued. 'Just two islands: St Agnes and Gugh – which you may already know are sometimes just one island, depending on tides.'

That was a bit deflating. There are around 40 islands with rats and this project worked on, let's be generous, just two of them.

'And that took until 2017,' Jaclyn said, unapologetically. 'Look, this work is far more than just scattering poison and hoping for the best. There is no point going off half-cocked, and that is why so much time was spent readying the ground, preparing the people who live on St Agnes for what was to come. We even got the experts from New Zealand over to give us advice – they really have set the standard for this sort of work.

'For example', she said, 'we needed people to really control their rubbish – reducing food options, and also not use rodenticide themselves for six months, so that when we started there was maximum bait take. Also, it was absolutely clear that there was no point using poison during the bird-breeding season, as the rats will simply gorge on eggs and chicks as normal. We needed to hit them when our offerings were at their most attractive. And this also meant that by being very selective as to when we put out the rodenticide, we end up using less – which everyone agrees is important.'

With around 85 human on the two islands, it was vital that all residents were on board with the project.

'It really has to be everyone,' she said. 'Not just those there all year long, but we needed access to holiday lets that could be standing empty too. They could be unused for six months, and could easily have a population of rodents gnawing on a stash of food. So this work, the time spent talking to people, was to get them to understand the importance of removing alternative food sources. And the best way to do that was to share with them what I had seen – Manx shearwaters, half eaten. These amazing birds – they use the stars to navigate around the world, and the UK has around 80 per cent of the world's breeding population.'

The combination of passion and storytelling is Jaclyn's superpower.

'We also had to be honest. When I started on this I felt I should not alarm people, but experience has shown the importance of being honest – about the time and money, and about the things we needed them to do, like not putting out bird food for the six months beforehand to make the bait seem all the more attractive – and warning them that there might be an increase in the number of rabbits as adult rats will predate kits.'

There were other consequences that were harder to predict. For a long time it was assumed that the shrews on Scilly were a distinct subspecies but careful measurements of skulls and teeth revealed they were actually the same as the lesser white-toothed shrews of the Channel Islands. Not that this makes them any less exciting. I rarely see shrews, but you can hear them chittering away on quiet summer days. I imagine that this will be an experience I will lose before I lose the screeling of swifts and after I lose the contact calls of bats, as my ability to hear the higher frequencies dulls with age. These shrews have benefited enormously from the loss of rats, but with an increase in shrews comes an increase in shrew-poo, so householders were advised to block even the smallest access points to their houses if they wanted to avoid a scattering of pellets.

Another beneficiary of the reduced rat population raises its own questions: pittosporum. Ironically, given the enormous amount of work New Zealand has put into trying to undo the ravages of imported wildlife, this evergreen shrub was deliberately introduced to Scilly from New Zealand. If there were to be a move to do the impossible and recreate Scilly in prehuman form, one of the most dramatic changes would be the loss of this plant.

Along with granite walls, pittosporum provides shelter on the island. I had noticed it on the walk the previous evening – there are complete tunnels of green deflecting the wind up and over the more sensitive plants. Amazingly, there are flourishing allotments in areas that without the pittosporum would be reduced to only wind- and salt-tolerant species. And it was for this quality it was introduced, to provide shelter for the bulbs that erupt into the heady fragrance that had already stolen my soul.

Rats eat pittosporum seeds. So the team working on the eradication had to introduce this potential change – that there could be a surge in growth to the point that the pittosporum itself became a problem. 'We have to be transparent as to the impacts, and also to the fact that many might not be anticipated.'

Shrews and pittosporum aside, the real issue is how the birds fared. Did the enormous amount of work, at a cost of £1million, make a difference?

In 2014, the year after the poison was applied, for the first time in living memory the burrow-nesting Manx shearwaters bred successfully. The following year the European storm-petrels returned. In 2022 there were 80 pairs of shearwaters.

But this is not the end of the story. All of the good work could be undone in one careless moment. One pregnant rat could generate a colony of 300 individuals in just eight months. When you arrive on St Agnes – which I failed to do as the boat to get us there was fully booked – you are greeted with a sign: 'Welcome to St Agnes and Gugh. We are rat free. If you think you've seen a rat please call the "Rat on a Rat" hotline.'

The human inhabitants have remained vigilant, and supported. Around the island, as on Lundy, there are bait stations with wax blocks inside – these are checked for teeth marks. So far, the biggest alarm was raised when a rat was seen jumping ashore. It was trapped, and it turns out that a delivery of wood had not been actively checked for stowaways. It is best practice to make sure that whatever is being shipped to a rat-free island is not only actively checked, but is also not left on the quayside.

Working with the RSPB project, Biosecurity for LIFE, Jaclyn has produced information for boat owners that gives simple steps to prevent inadvertent rodent trafficking. So, if you find a rat or mouse while you are out in the boat, do not land on a rodent-free island, return to port and get rid of it. There are guards you can attach to mooring ropes that prevent the gymnastically inclined mammals from reaching the boat – they are a smaller version of the baffles used to try and keep grey squirrels off the bird feeders. Leaving wax blocks or 'chew cards' around the boat will allow you to see if you have unwelcome visitors.

There is nothing really complicated or surprising about the guide. Making sure that food on boats is stored in rat-proof boxes, that bedding is shaken out away from the boat, and ensuring that potential entry points are blocked – these are all things that anyone using a boat would be likely to do. It is just providing a checklist with the emphasis on biosecurity.

Probably the most important part of the follow-on work on St Agnes is the 'incursion shed'. This contains all that is needed should there be a potential sighting (monitoring equipment and maps) and, as with Lundy, there is the potential to leap into action with bait stations.

'One way I can tell that we have done well with this,' Jaclyn said with obvious pride, 'is how proud the community are to have a rat-free island. Primary school children are as involved as farmers and fisherfolk.'

Two islands have been worked on, but surely Jaclyn is not going to stop there?

'There has been a bit of an interruption,' she said. 'After we were sure that the islands were clear – and we had to wait for two years to be able to declare a success – Tristan and I got married.'

Obviously I offered congratulations. So was it a particularly long honeymoon getting in the way of the next step on Scilly?

'Not really a honeymoon, no,' she said. 'I got offered another job that would take me away for six months – all the way to Lord Howe Island.'

This Australian island – or more precisely, island group, as there are 28 islands, islets and rocks – is 600km east of New South Wales, and forms a 10km-long crescent created by a long-extinct volcano. The richness of the biodiversity helped the island earn a place on the list of World Heritage Sites, and it is now being actively protected under the auspices of the New South Wales government.

It certainly needs protecting. It was sighted and named 1788 by British lieutenant Henry Lidgbird Ball. Settlers arrived in 1834 and it became known as a place for whalers to top up on food and water. With the demise of the whales, the export of the native kentia palm as an ornamental plant became the main industry. The first tourists arrived, with the advent of a regular shipping service, in 1893, and now tourism is a mainstay of the island economy.

Inevitably the arrival of people – there are around 350 now – brought other species, species that have caused considerable problems for the native wildlife. Lord Howe Island has undertaken really serious programmes to control the most damaging of these new species, and it is clear from Jaclyn that this level of willingness to invest resources into the work was a good reason to drag her away from her new husband for six months to finish off the work on the Lord Howe Island rodents. After all, she is now a go-to person for this sort of thing.

The biodiversity management plan for the island, produced in 2007 by the Department of Planning and Environment, gives a fascinating insight into some of the new arrivals. There is no generalisation, no 'foreigners out' politicking. In fact, the attitude within the plan is really balanced, saying, 'The degree of impact, and interaction with native and endemic species, needs to be assessed in order to determine which introduced species may warrant control programs.'

For example, both blackbirds and song thrushes were introduced in 1944 – not just because their beautiful song enlivens a garden, but with the thought that they would control the weevils that were eating palm flowers, thus reducing yields. The birds are still to be found on the island, and they do eat endemic invertebrates, including the endangered Lord Howe placostylus, a large land snail. But the biggest threat to this mollusc comes from rats, so the birds are not in the conservationists' sights.

Masked owls were introduced in the 1920s in an attempt to control ship rats – you can see a pattern developing here. Obviously that did not work and they are now considered a pest, preying not just on unwanted rodents but also on white terns, Black-winged and Providence petrels, Lord Howe woodhens, and endemic skinks and geckos. It is possible that they are preventing Kermadec and white-bellied storm-petrels from recolonising the main island as well. Again, as with the blackbirds and thrushes, while the owls are part of the problem, and are being controlled, their impact is not as great as that of other incoming species, so that is where attention is turned.

One of the most effective destroyers of island life is the domestic cat. These arrived in the 1840s and had a predictable impact on native birds and reptiles. Action was taken, and by 1979 all feral cats had been removed. Immediately the Lord Howe woodhen benefited, along with the wedge-tailed and little shearwaters.

Feral cats are all well and good, but what about domestic ones? This is where it can get tricky. In 1982 the government introduced a local law banning them, but with a grandfather clause that allowed owners to keep their existing cats as long as they were neutered.

The bleating tree frog is a more recent introduction, though the questions of who introduced it and why remain unanswered. Its distinctive call is common, but the only impact it is thought to have is as a competitor and predator of some of the local invertebrates.

Around 5 per cent of the 2,000 or so invertebrates on the islands are introduced, and of these it is the African big-headed ant that causes most concern. Ants are no joke – the rather brilliantly named yellow crazy ant has made a new home on Christmas Island and in the process has killed 20–25 per cent of the entire population of land crabs in seven years by spraying crabs encroaching on its territory with formic acid, rendering them blind and immobile.

Invasive plant species are causing some of the biggest problems by competing with native species, and have the potential to transform entire island ecosystems. With more than 670 invasive species on Lord Howe Island, there is plenty of work there too.

But this is never going to be a case of rediscovering the prelapsarian idyll. The work that has been undertaken has been to identify the biggest threats and act on them – and perhaps in time learn to live with what remains.

The biggest threat was, predictably, the remaining rodents. Mostly rats. Rats have been linked to the extinction of 5 endemic bird, 13 invertebrate and 2 plant species on the islands. Rodents are recognised as a threat to at least a further 13 bird species, 2 reptiles, 51 plants and 7 more invertebrate species.

The job that Jaclyn was recruited to undertake was far from simple. 'The rodent project had been rumbling on for around 15 years when I arrived to finish it off,' she explained. 'The problem was the divisions within the community. And bizarrely part of the conflict resolution came down to religion.'

Jaclyn is blessed with many fine qualities, including boundless enthusiasm. But it is her parents that inadvertently helped her in the situation she was confronted with on the islands. 'My parents were accountants, and my late mother in particular was very religious, so I was brought up with an understanding of the deep-seated nature of belief. The reason this was important is simple. There was no point starting the final part of the rodent project on the islands without the buy-in of everyone. This was not an attempt to control a problem. This was an exercise in eradication. If we leave any behind, all the work, the time, the money is wasted. And for this to happen I needed to have everyone working with me for the final goal.'

The problem she faced came from the complexities of religions on the island, with Catholics and Protestants communities on the island and also Seventh-day Adventists who maintain a belief in the 'infallibility of scripture'.

'I started my six months by listening, trying to identify the problem,' said Jaclyn. 'Initially I was using community meetings, but quickly realised that the people who turn up to these meetings are either very in favour of the plan, or very against it; you don't get a broad cross-section of opinion. So I resorted to going door to door, making calls.'

It turned out that the obstacle was that members of the Seventh-day crowd believed the presence of the rodents was God's work and therefore they should be left alone. If God wanted them gone, then God would get them gone.

'So I started going to church every week – my religious upbringing made this easier – and over time, I got to realise that we were looking at the island with a completely different set of values.'

It is very easy to assume a shared point of view, and it can be quite destabilising to be reminded that this is not the case. I remember while radio-tracking hedgehogs in Devon in the early 1990s going for dinner at the pub and realising that my bubble of a vegan life in Manchester was very small. I ended up with very lovely salad and chips, but it was a moment I don't forget.

'I took people out to see the birds that were being predated by rats,' Jaclyn said. 'I assumed that this would overcome their suspicion of the trail-camera footage. What I was not expecting was "I genuinely don't care." How can I work with that?'

'It turned out that what motivated them was their phobias, and the damage the rodents caused to them personally. But the real turning point came after around a year when –'

I had to interject. She had just got married and had only agreed to leave Scilly for six months.

'Yes, that was an awkward call. In fact I ended up spending two years on the islands … but I am back now and we are building our forever home. Back to the turning point – well, it was as simple as the congregation recognising that perhaps my presence on the island was God's work, and that my being there was evidence of the need to remove the rodents.'

The necessity for everyone to work together was clear. Because for this to work there was a strict regime that needed to be followed scrupulously. 'As with St Agnes and Gugh, we needed six months with a complete change of behaviour from everyone,' Jaclyn explained. 'I could not just arrive and dump poison. Rodent eradication is not that simple. We needed to ensure that there was maximum impact, and that meant setting the ground.'

Setting the ground required that people stopped putting down their own poison and throwing scraps of food out into the wilder patches beyond their gardens, so that when the poison arrived it would face less competition. And she needed access to everyone's home as well, so this could not be done without complete support.

Winning the hearts and minds of the people of Lord Howe Island was the hardest bit of the project. When it came to the physical eradication, that was relatively straightforward, taking place between May and November 2019, and representing the culmination of this AU$16-million programme.

More than 130 tonnes of the cereal-based bait Pestoff 20R – which contains 0.02g/kg of the anticoagulant poison brodifacoum – were distributed around the island. Bait stations were placed in areas around livestock and settlements, while in the uninhabited wilder reaches of the island, helicopters dropped bait from a special spreader. It would be great to rely on bait stations, as this reduces the incidence of accidental poisoning, but the scale of the project was just too vast.

'We did it,' Jaclyn said. 'We managed to kill off all 300,000 or so rodents on the island. And there is no way this could have been done without the community supporting us. Also, I do recognise that slathering an island in poison is not ideal. There will be knock-on impacts, some non-target individuals will have been killed. But what was the alternative?'

That is a really good point – there is a tendency to let the perfect be the enemy of the good. And sometimes we have to accept that good is, well, good enough. In years to come there may be a more species-specific poison or a perfect lure to a trap, but if we wait long enough for perfection, we risk those species at risk being driven to extinction.

Just as we were wrapping up this fascinating conversation Jaclyn mentioned something that happened on Lord Howe Island that must be among her worst nightmares. 'In 2021 two rats were spotted on a road. Within two days they had been trapped and killed – the female was pregnant. There is a rapid-response team there, just as we have here on Scilly. These animals are amazing – they continue to confound us humans. We should never underestimate their ability to thrive where we would rather they did not.'

She did eventually make it back to Scilly, and is already building towards the next project. 'My parents were really confused with my work,' Jaclyn said. 'I was the one rescuing spiders out of the bath at home, now my job is killing rats.'

This time she is aiming at removing the rats from three inhabited islands: Bryher, Tresco and St Martin's. This is still early stages, and has added challenges: more people and complicated land ownership, as Tresco is privately owned. But it won't just be her. 'We have a partnership of the Isles of Scilly Wildlife Trust, RSPB, Natural England, Lucy of Cornwall, the Scilly Area of Outstanding Natural Beauty, along with representatives from the islands. Just because we have made this work once, we can't get arrogant and just step in there.'

Jaclyn did repeat almost word for word the sentiment of Rosie on Lundy. She too has nightmares about returning rodents, and she too sees her work to be keeping them at bay. Reasons for constant vigilance are plentiful.

'If you want alarming, have a look at this information from the RSPB biosecurity manual.' Jaclyn showed me another image on her phone. It had the distances many of the less-welcome species can swim. Brown rat, black rat, mouse, stoat, mink ... just take a moment to think how far an island would need to be from a potential source. Respectively, the maximum distances these animals have been recorded swimming is 4km, 750m, 500m, 3km, 4km.

Though crossing a distance of 50m or less is considered an easy task, this does not mean that islands within smaller distances are inevitably going to be invaded. Scilly is blessed in part by some fairly vigorous currents, making the journey harder, but still, there is no room for complacency, and the researchers who came up with these swimming figures point out that they will only ever be revised upwards.

And what about hedgehogs, I wondered? The bus driver from the airport told us stories of seeing hedgehogs and helping them out of the way – this happens to me a lot. People ask what I am doing, I mention hedgehogs and then follows – more often than not – anecdotes.

'We love hedgehogs, there is a woman in Hugh Town who rescues them,' Jaclyn said with customary brightness – I was looking for the but-face ... yep, there it was – 'But, only while they are here, on St Mary's. You see we have so many other problems threatening the birds here that there is no point trying to remove hedgehogs. However, if hedgehogs were to get onto some of the other islands – which would only happen if someone thought it a bright idea and popped some over on the ferry – then it would be a problem, we would have to act and remove them. You know why more than most people I imagine. In fact, would you help in the campaign?'

Since then the British Hedgehog Preservation Society, of which I am the spokesperson, has stepped up to support this work. We know, as Jaclyn pointed out, all too well what can happen.

'In fact, before you go, could you have a look at this? she said, reaching for her phone. 'I was sent this a few days ago; the person who took the photo thought it might be a hedgehog footprint, from Gugh. We have already put out some cat food and a trail camera, and there is a "Hedgehog Scare" WhatsApp group. But as you are here, what do you think?'

I am not the world's best tracker. I have friends who are deeply skilled at bushcraft; mine is a limited range. And while I am usually delighted to be able to tell someone they have shown me a hedgehog footprint, in this case, I was just delighted to be able to tell then that it was definitely not a hedgehog footprint. Looked more like cat to me ... and which would be worse for the birds? Now there is a question.

It was Jaclyn's passion and her good nature that were as important as poison in getting rid of those rats. We could have carried on chatting for the rest of the day, but Jaclyn had work to do and then there was the matter of the quiz night.

Miriam reappeared from her Neolithic adventure and we started to plot another walk for the afternoon, which again took us into the dark. This time our route took us through the churchyard. I am sure you know when things feel a little different. Well, they did here. I could not quite put my finger on it, no, nothing 'woo', this was much more mundane. I could not place the trees, bare of leaves. Large, mature and just different. Miriam had the answer – these were elms, proper, adult elms. That generated quite a wave of emotion. I was too young to notice their demise due to Dutch Elm Disease on the mainland. Here, they have escaped. How precious these island sanctuaries are. Or, as Miriam put it in her article she wrote for the *Times*, 'To come here is to step into gentle-time, a stolen moment, a secret garden.'

Galapagos

I am not proud of the feelings of jealousy that poured over me as I read Emma Marris's *Wild Souls*. It is a beautifully written, wise book about where animals fit into this human-dominated world. The jealousy was not for the lovely words – writing books is not a pie – I want people to be brilliant so I can enjoy them.

No, the green tendrils crept into my heart because of the opportunities she has had as a writer for *National Geographic* and the *New York Times*. She has been to all the places that I have dreamt of visiting. And one of the stories she told is of that group of islands that is at the heart of our modern understanding of life on earth – the Galapagos.

Like we have on all islands we humans have visited, we have left our mark. For example the Indefatigable Galapagos mouse was wiped out by human-brought rats on Santa Cruz Island. The mouse was called Indefatigable, not because of its ability to ride out any disasters – which would have been rather ironic given its extinction – but due to the previous name for the island, itself named after a ship.

The Galapagos penguin has many reasons for its fame: the third smallest of the penguins; the only one to live in the northern hemisphere; and – thanks in part to a single cat that was eating around seven every month before it was dispatched – the rarest penguin on the planet, with fewer than 1,000 breeding pairs left.

Marris got to spend time with Karl Campbell, who works for Island Conservation, an international NGO that researches ways of removing invasive species. They tend towards the euphemistic, referring to the distribution of 'conservation bait', but are at the cutting edge of developing drone technology to deliver this poison precisely where needed.

Karl is keen to use this conservation bait on the island of Floreana to assist in the return of the Floreana tortoise – or a creature as close to that extinct species as possible. What he has found are hybrids. But they cannot be released until the rats are removed, as they eat tortoise eggs and babies.

The Galapagos suffers from more than just rodents and cats. One of the most concerning species is the accidentally introduced parasitic fly, *Philornis downsi*, which is rather dramatically known as the avian vampire fly. About the size of a housefly, it lays eggs in the nests of birds; the eggs hatch into maggots that feed on the nestlings. This has increased the mortality of the famous Darwin finches, leading to a decline in the population of many species of these birds. So far there has been no success in finding ways of removing these insects from the islands.

As an aside, I imagine that most people would not be worried about the ethics of killing avian vampire flies whose maggots feast on living chicks. We would be worried about the knock-on effects if poison was used, I am sure, but we have to admit to a degree of straightforward species preference – speciesism, as Peter Singer describes it. I don't think there is anything wrong with that, but I do think it is important that we recognise that killing for conservation is not an absolute right or wrong. Even the most ardent normative thinker from the world of animal rights would be hard pressed to argue against killing this fly.

Also presenting a struggle on the islands is the smooth-billed ani, a sort of cuckoo that was introduced in the 1960s from South America by farmers who had seen it feeding on cattle ticks. Ticks had recently been introduced to the islands by accident and were causing cattle to suffer and die. So, what better way to control the ticks than their natural predator …

But the anis ignored the ticks, and have instead made themselves at home robbing the nests of endemic mockingbirds and Darwin finches, as well as native yellow warblers. They have also targeted the Galapagos lava lizards and large numbers of endemic insects, including the only native bee on the islands. If that were not bad enough, they are also brilliant at dispersing the seeds of the invasive blackberry, a bramble that is causing enormous ecological damage.

There is as yet no effective eradication programme in place for this out-of-place bird, and because this is ecology, there is another twist to the story. The smooth-billed ani is also a predator of a wide range of other invasive species. Mice, rats, wasps, cockroaches are all targeted by this bird, and if that pressure were to be removed, there could be a surge in their numbers.

Back to Karl Campbell. His grand plan is to saturate Floreana with brodifacoum to eradicate the rats, but in such a manner that farm animals and native wildlife are not killed. This will be the largest and most complex eradication programme undertaken on a tropical island. Island Conservation aims to restore 40 island ecosystems by 2030, and they have set in place systems to ensure damage is limited. For example, they will buy 'sentinel pigs' – sacrificial pigs to be more honest – who will live in the areas being poisoned and then slaughtered to check for toxin build-up. Until they are clear, the resident pigs will have to be kept enclosed.

Native birds are being caught and kept in aviaries, on and off the island, to be returned when the island is free from rats. The scale of the project is daunting: $26 million over 10 years. And if it is not 100 per cent successful, well, it has failed.

CHAPTER 6

Orkney

Spud is one of the most handsome dogs I have ever met. A sleek black Labrador with a strong work ethic, equipped with a remarkable 'scat-nav'.

Spud, and Stuart, his human, work for the Orkney Native Wildlife Project (ONWP) and it took me a very long time to manage to get to meet them. Illness, a clash of dates with a toad summit at which I was keynote speaker, and logistics, all conspired. But I got there in February 2023, and managed to leave a little time for myself. My meeting with ONWP boss Sarah Sankey was to be the day after I arrived, giving me half a day to play.

Orkney has a very powerful hold over me – not all positive, but certainly strong. It is where I did my first-ever hedgehog research back in 1986 up on North Ronaldsay, looking into the impact that imported hedgehogs were having on the breeding success of ground-nesting birds. I have been interested in this subject for a long time.

That visit, with a colleague from Leicester Polytechnic, Kerry Seal, was instrumental in me getting to where I am now. And it is a disappointment that I lost contact with Kerry so if anyone knows where she is now, I would love to get in touch. We took a very long time to get to Orkney. I remember calculating that we could have flown to Australia in less time than it took us to get to the mainland of the archipelago, never mind the stomach-churning journey on the ferry to our tiny destination.

We did take a day to ourselves on arriving in Kirkwall and jumped on a guided wildlife and archaeology tour bus. That is when the seeds of love for the islands were sown, and I got a chance to revive them this time around.

I was conflicted – should I go and see the most glamorous bits of Neolithic standing stones first, and then head to the coast to watch the sun set into the Atlantic, as it was a beautiful day? Or see the sea and photograph the stones in the setting sun? While buying supplies for my stay – bed and breakfast these days tends to be less generous on the breakfast – I decided and headed west to the coast.

Near the Neolithic village of Skara Brae is the Bay of Skaill. Parking at a small kirk, I started walking around the edge, soon leaving the path to get down onto the slippery rocks and closer to the elemental force, looking to where the waves come from. There was nothing until Newfoundland. A fulmar gently fell into the air to float in a serene swoop around me – I remember those beautiful eyes, gauging whether I was a threat. Mallimacks, they call them here. I had remembered it as being mollymawks (also a name for some albatrosses), but that could be down to pronunciation.

The sea was all I had hoped it would be – thumping with power, sun shining through the waves so they glowed mint green as they reared up, and the horses – there really were horses in these waves.

Sated and salted, and before the sun had set, I headed to the stones.

The Ring of Brodgar is the most complete of the circles on the islands and is rightly famous. Since I had last been up there, the amazing discovery of the Ness of Brodgar had been made public – a vast Neolithic development that is being slowly and carefully excavated. It is thought to have been a spiritual hub, rather than a functional place like Skara Brae. Walking around the stone circle, and not for the first time, I found myself wondering whether archaeology might not have been an interesting alternative to ecology. Too late now.

I could not get to the stones; they were fenced off and I was somewhat disheartened by this. Yes, I understood that it was to protect the area from erosion, but I had travelled a very long way to get my moment of connection with these amazing relics. I waited until no one was around and stepped over the line to take a moment with them.

It is a bit like our relationship with nature – when somewhere becomes too popular it needs to be controlled, or rather, we need to be controlled from damaging it. This is the reason that the title for my first book was supposed to be *The Hedgehog's Dilemma* – the Schopenhauer idea that if you get too close to the one you love you end up causing pain. We need to find the closeness that brings connection without causing damage, pain or suffering.

I moved on to the nearby Stones of Stenness, which is not as complete as Brodgar, but accessible. And as for the sunset, well – I really had a moment. In between taking photographs that would happily adorn adverts for the local whisky, Highland Park, I took time to stand and just think. Mainly about my friend, the comedian Robin Ince. He wrote a beautiful poem about ghosts – which for someone driven by empiricism is hard to imagine. But he was captured by an experience with a standing stone on Lewis, in the Outer Hebrides. The concluding line is a delight:

Emergent complexity briefly defeats the void.

Strange how the great expanse of sky and nature and time can lead one inwards. As the sun set, I was visited by a melancholy that seemed very fitting.

The 3.00 a.m. start was catching up on me so I drove to my room, ate food and processed photographs before slipping into an easy sleep with the windows and curtains open to the wild world just outside. The stars were bright, the air crisp, and the wind soothing.

The Orkney Native Wildlife Project is based in a Kirkwall industrial estate – rather an unprepossessing warehouse amid tool hire and car repair companies. The poster on the door, 'Seen a Stoat?', was a clear indication that I had come to the right place. For it is stoats that Spud, five other dogs and 36 humans are dedicating their lives towards.

Sarah met me with a ready smile, introducing me to her colleague Lianne Sinclair. As we settled down around a table and I started to ask questions, Sarah interrupted me and said that she had arranged things a little back to front, and that we were heading out on the road now as the weather was good and she needed to check a site on South Ronaldsay. Then we could have a conversation when we got back.

This was in large part due to the hangover of good weather that was treating Orkney to a third day of sunshine, and the forecast suggested this would not last. While waiting for the team to assemble, I was shown into one of the more functional rooms – stacks of wooden boxes, designed to take traps, and a huge box of the snap-traps that are the sharp end of this business.

Sarah and Lianne joined me in my car and we headed south, pretty much as far south as you can go, which was only 20 miles from Kirkwall. This meant crossing the Churchill Barriers, built between the islands during the Second World War to block access to the anchorage of Scapa Flow. They now carry the road that connects mainland Orkney with Burray, Lamb Holm and Glimps Holm – and this is significant for this story.

We were on a quest for stoats. These amazing mustelids are a really new addition to the fauna of these islands; the first records come from just 2011. 'They were found in two different locations around 30 miles apart,' Sarah explained. 'And we still have no idea how they got here.'

As is the case with so many islands, a new predator can cause trouble. Orkney is home to many ground-nesting birds and any visitors will know why – trees are thin on the ground and those that have managed to take root tend to be heavily leant on by the wind. Orkney also has its own vole – a subspecies of the common vole. The common vole is really not very common in the United Kingdom, in fact it does not live there at all, with its home being all over the continent.

How Orkney's vole got to Orkney has been a matter for debate. The original theory is that it was a relict population from when there was still a land bridge between the islands and Scotland. Now it is believed they arrived during the Neolithic, while people were building villages and erecting stones. The oldest radiocarbon-dated specimen is from around 4,600 years ago. Probably they came hidden in animal feed, as the Celtic highway from the continent, around Ireland and up to Orkney was a far more active route of travel than a trudge up through Britain.

'Our voles are precious to us,' Lianne said, 'And really quite chunky.' She was not joking; they are twice as heavy as their continental cousins.

As for the stoats, their journey could also have been made while hidden in bundles of animal feed or stashed away among crates on the ferry. But they are not like rats or voles – they just don't tend to do that sort of thing. It is possible that someone thought it might be a good idea to try and control rabbits so brought some over, but really, with what we know now about the implications of introducing new predatory species to an island ecosystem, that takes a very special sort of stupid.

With the sightings of stoats the RSPB reacted fast. 'There can never be a period of stasis,' Sarah said. 'Every day they are multiplying, every day could be the one in which they swim to a new island. So we got a volunteer effort going, which worked quite well to begin with. Where we are going on South Ronaldsay was once the epicentre of sightings. For a couple of years, they seemed to be being kept under control, but the reality was that within seven years all 227 square miles of the mainland and connected islands was covered.'

Stoats can swim for a bit, but the Churchill Barriers made progress far easier.

'It is when you look at the breeding habits of the stoats that you understand how this happened,' Sarah continued. 'All the females are pregnant pretty much all of the time. As soon as one gives birth the male is on her, impregnating her again. But he does not stop there, he also mates with all the female kits too, even if they are his daughters. Now, they are too small to go through pregnancy, but they have delayed implantation, so as soon as they are ready, there is no need to find a male.'

The first place they went for advice was New Zealand, as stoats are on their hit list too. The message came back loud and clear: 'Act now'. The problem on Orkney could be even worse than in New Zealand. There the food fluctuates, but here, while the breeding birds come and go, the voles are always present.

But acting now does require a bit of a run-up. 'As we were bidding for public money we had to undertake due diligence,' Sarah explained. 'So after establishing this was a real threat, we needed to undertake a technical survey. And while that was going on we instigated a biosecurity programme – trapping along the coast where there was a chance of them swimming to another island. Having them on Mainland was bad enough, but if they spread to other islands, that would be disastrous.'

That trapping was in 2016 and around 250 stoats were killed.

Once a technical survey had been done, and before funding could be secured, there was another step, one that is becoming increasingly clear as an essential component of any such wildlife management. And that is to assess the social feasibility of the planned work.

There was a time when conservationists would 'know' they were right and step into a situation with little care for the feelings of people who were already living there. As we have seen on Scilly, without the buy-in of the local community, these projects just don't work. And the same was true on Orkney.

'Probably the hardest part of this job has been communication,' Sarah said. 'In the first instance we sent out thousands of questionnaires, and did so many talks. But doing an event tends to only attract people with strong feelings, and mostly those who agree with you. So we also had to go knocking on doors. There are around 19,000 people living on Mainland and the connected islands.'

'We would miss some like this,' Lianne added. 'So one of the ways we could reach new people was to follow the mobile library that travels around the islands. This was great as we would pitch up a small stall next to it and be able to reach new people. With that, and the school visits, we got really good coverage and the overwhelming result was that people supported the idea of removing the stoats from the islands.'

'But general support was not enough,' continued Sarah. 'We needed access to the land, so worked on access agreements with landowners. There are over 800 of these now, and most were done in person.'

This is very different to an almost empty island like Lundy, where the only people to talk with are part of the same project.

Money was raised, a team collected, staff recruited, access agreements sorted, traps ordered, and work began at the start of 2019. The scale of the project is ambitious and a little daunting. More than 6,000 traps were bought from New Zealand. These were the DOC 200s I had seen in the warehouse. They are a vicious snapping device with a weight-adjustable treadle, so animals under 100g will not set them off. These traps are then set inside a wooden box over 1m long, partitioned with thick mesh set to allow in only stoat-sized beasties. Smaller animals can get in, of course, but the scales are set in their favour. Anything heavier triggers the spring. Nothing has been found injured – death is sure and quick.

The first traps went out in September 2019 around the two areas where most access had been agreed and where stoat presence was high. One of those areas was where we were heading now. Every now and then I would notice a roadside trap; they had been thorough.

'It started so well,' Sarah said. 'Traps were out, every 250 metres, and we had a routine; a brilliant app on our phones which meant that every trap was recorded, and when it was checked, every three weeks, the results reported. We were building up so much good data about the spread of the animals and where the traps were best sited.'

But then – the islands were not immune to Covid, and restrictions on movement brought the project to a juddering halt. 'All of the first six months' work was undone,' Sarah sighed. 'And while I did what we all did, and resorted to Zoom for meetings, the stoats were breeding.'

Undaunted, the team returned to the field as soon as they were allowed. In fact, it was not so much the fields, as the results were showing that the stoats tended to avoid the farmed landscape, concentrating their interests around the coast and in the tussocky, rougher places.

I asked about how much they hated stoats (okay, this was mischievous). 'The reason for this project is to protect wildlife, not to kill stoats,' Sarah was quick to respond. 'If there was another way, believe me, I would embrace it. No normal person is going to enjoy killing wildlife. These animals are amazing – intelligent, resilient, sentient – they deserve respect and are only being killed because we humans did something stupid.'

As we arrived at the southern tip of Orkney, Stuart was already there, getting Spud out of the van. He had to go for a play first – there is clear demarcation between work and play – so Sarah and Lianne walked me up to what must be the most unlikely picnic bench ever, utterly blasted by the wind on what was really quite a nice day. 'Never seen anyone using that,' Sarah said when I suggested there must be quite a hardy bunch up here.

Soon Spud joined us, and Stuart explained that there had been a stoat caught in a trap down here two weeks ago, so now they needed to see if there was any other evidence. This is where Spud came into his own – his sensitive nose can pick up the trails that stoat have used a couple of days ago, revealing activity. How does Stuart know when Spud has hit on a trail, I wondered. 'Oh, you get used to reading him,' he said. 'The main indication is his tail – it goes into full helicopter mode.' And with that, he leant down to Spud and put on an orange harness, telling us that he was now an ONWP Conservation Dog, and telling Spud that it was time for work.

Then Stuart simply said 'search', and Spud was off. We hung back as Stuart followed the dog, quartering the tussocky grass. Now, in the hedgehog world we have a conservation dog too – his name is Henry and while I think he was once into drugs, now he works on finding hedgehogs, which is a crucial step on the way to helping developers make good choices. Hedgehogs do not have the amount of protection that obligates developers to check an area for their presence, and on top of that it is actually quite tricky to find them. But Henry is great. However, he can only work in short bursts, needing a breather every 15 minutes or so. This is not the case with Spud. Turns out he can outlast his humans by quite a chunk of time, especially when they are laden, as Stuart was, with two emergency traps and all sorts of other necessary gubbins.

After about 20 minutes Stuart was convinced that this area was clear. He found one of the thousands of traps had been disturbed and was initially worried it could be the work of a disgruntled human – which I had the temerity to question. I remember when I would find my small mammal traps – which caught mice and voles alive, allowing me to monitor populations across different ages of coppice – started to be broken open, releasing any potential rodents (and data). I had thought it might be a misguided animal rights activist, so I hid out in the woods in a raised hut used for shooting deer, and waited, and waited, until a squirrel turned up, and moved along the trap line, opening each one, eating the seed that was in there as bait, and then onto the next.

Stuart looked more closely and came to the conclusion that it was probably a nosy otter. They would not turn their noses up at the bit of rabbit used as bait, but – unless they came equipped with power tools – would find gaining access quite tricky.

This was just the start of the outing. Another trail dog had picked up a scat further around the coast, so we followed on behind Spud, now on a lead. I could see why – he was so focused on following his nose that he ran perilously close to the cliff edge.

Even if there was not the fascinating company and the brilliant dog, this would have been just a perfect walk. The wind was whipping up more white horses, some of whom were crashing pleasingly into the cliff. I had forgotten the deeply cleansing feel of this wind, tangy with the scent of sea and kelp. Lianne saw me looking out and breathing deeply, 'It is the thing I always realise I have missed,' she said. 'The smell of the air here is unlike anything you can find back down south.'

Lianne was born and bred in South Ronaldsay and carried with her the wonderfully soft accent that I remember from previous visits. Sarah and Stuart are incomers, though with long standing. 'Oh,' Lianne said, 'I speak "properly" when at work – it all goes a bit stronger when I am with family.' And that too I remember from North Ronaldsay, some of the older folk in particular, were difficult enough to understand when sober, and by the time a few cans of Special Brew had been consumed, smiling and nodding was the best I could manage.

Damn, that moment of chatting had taken our eyes off Spud who was now being given love and praise. We dashed, well, trudged onwards towards the happy dog. Oh, the tussocky grass, I had forgotten how hard it was to traverse, and what an amazing habitat it made for stoats and voles (and a shrew, which I spotted briefly). By the time we got to him, Stuart was reaching into his pocket – for a treat I assumed, but no, this was special. Spud had found a scat – a stoat dropping – and for that he got his ball. The absolute glee on his face as he munched the orange and blue ball was only topped by the treat he got before it was time to get back to work.

And Spud certainly deserved that treat. Sarah described how it was found; they were walking into the wind when Spud pulled Stuart 20m downwind and straight to it. It makes my mind boggle to try and grasp the way in which other animals 'see' the world.

Stuart showed me the scat, like a very small fox dropping. He had already collected it into a glass jar, careful not to touch it, to keep the scent from his fingers. Every one found was taken back and analysed. There is much to be learned from mammal poo, including – crucially – who the animal is. If 20 scats are found and they are all different, you know you have a very big problem. Whereas if it is just one individual that has left them all, well, it may not be a time for complacency, but it is an indication of what you are dealing with. However, there was no real concern about this finding, which I found surprising. 'It is really quite old,' Stuart explained. 'And Spud has found no other signs, so this leaves me confident that the one that was here has been removed.'

'Look, I am sorry to drag you away', said Sarah, 'but we have a meeting about budgets at 3.00 p.m., and it would be good to sit down and talk through the numbers of what we have achieved so far.' So with that I took my leave of Spud and Stuart and we began our journey back to Kirkwall.

Sarah had come to the island to work as a wildlife tour guide, but before then had studied in Oxford and worked for the local Wildlife Trust, all while I was around the city too. Shame our paths had not crossed earlier. Lianne had left the islands, occasionally and briefly. I really got the feeling that her heart was tied closely to this land. And I could see why – the lack of clutter leaves the mind freer to think. It is no surprise that Orkney punches well above its weight in terms of artists, poets and musicians. Orkney's own Peter Maxwell Davies is on my ever-increasing list of people I wish I had met.

The drive back north was companionable, and by the time we got to Kirkwall we were all hungry. 'Have you tried a macaroni cheese pie yet?' Lianne asked. While I understood each of the words, I could not quite understand the concept as a whole, so clearly I had to give it a go. There were just two left at the amazingly well furnished delicatessen near their office – Sarah grabbed one, while I took the last.

Sitting back in the office Sarah said, 'There is no tidy way to do this,' and bit into the small pie. A pie of maybe 5cm diameter with a high wall of crisp pastry held the prize, macaroni cheese. There is no way that this should work, but it does. And there is no way of eating it without making a mess – I possibly went down in their estimation when I went to find a knife and fork. Prissy southerner!

After making short work of the surprisingly wonderful discovery, it was back to finding out how this project was really going. Because it is remarkably ambitious. 'This is the first time anyone has ever tried to remove stoats from such a large and populated area,' Sarah said proudly. 'And it is the first stoat eradication attempted in the northern hemisphere.'

The £8-million budget, secured from a range of sources – the National Lottery Heritage Fund, the EU LIFE project, the RSPB, NatureScot – is huge for wildlife eradication work in the United Kingdom. How long was it going to take, I asked Sarah. 'Well, initially we bid for five years – one year setting up, two years knock-down, two years mop-up and then two years monitoring.'

Even I could spot the problem with her arithmetic. And that was without adding in the lost season from Covid. But Sarah knew this was never going to be simple. 'There is such time pressure,' she said. 'Every day the problem is just getting worse, every day we leave it, more stoats can be born, making our work harder. And then there is the nightmare of one getting to Hoy, for example.'

Hoy is the most mountainous of the archipelago, famous for the 137m-high stack, the Old Man, that greets the ferry. And if the mustelids were to make it here, the task of eradication would be made dramatically more difficult.

Echoing Rosie and Jaclyn, Sarah said, 'These are the things that fill my dreams, and wake me in a cold sweat. I am, our entire team are, dedicated to the eradication of stoats from Orkney.'

But stoats are not the biggest concern of many of the landowners that Sarah has had to work with to ensure access was allowed. Their worry is the damage caused by a bird – the greylag goose. And as she is from the RSPB, this must be her responsibility, they assume. But it is, like everything ecological, complicated.

The geese have always visited the islands, and with the climate changing, so have their behaviours. Around 30,000 are now resident, with a further 40,000 stopping over during the winter. Farmers are losing crops to flocks of geese and are wanting a solution to the problem. Licensing has been eased on shooting, but even so, with flocks this size, unless farmers are very well organised, little will change. Some farmers have gone as far as altering which crops they grow, to avoid planting goose-fodder.

'I completely understand their frustration,' Sarah said. 'But there is very little I can do. Perhaps the most important thing I do is show them quite how much work would be needed to make a difference – and for that I use our stoat project as the example. I know that some of the farmers would like to use the geese as a bargaining chip, but it is just not as simple as that.

'We are lucky that Orkney has always been such a hotspot for not just wildlife, but also for enthusiasts,' she said. 'So we have decades of records for the voles, skylarks, meadow pipits, seabirds, hen harriers etc, so we have been able to see how populations change over time and will also be able to see the difference the lack of stoats make.'

They have been so busy – more than 4,300 stoats had been caught by April 2023. Though the alarm I get from that is how quickly the population had grown, with the first sightings being only 12 years ago. The results are beginning to be seen, with an increase in signs of voles being the most obvious, and some wading bird productivity improving.

There is one contentious issue that has not come up. When I was up on North Ronaldsay looking at the threats the ground-nesting birds face, it was clear that hedgehogs were part of the problem. However, there were also a great number of feral and not so feral cats roaming the island. Cats are another introduced predator with a reputation for wreaking havoc. The question is what do you do about cats … which is what I put to Sarah.

'Of course cats are a problem, for all the same reasons that stoats are a problem. But remember, this project could only work because we undertook a social feasibility study alongside the technical. We would have to do the same if we were to try and remove cats, and we just know that it would fail to get public support. And without public support, there is no point trying.

'If we were to try and remove just the feral cats, well, that wouldn't work either. It is now illegal to bring hedgehogs, stoats and other predators to the islands. So we have largely cut off the source of these animals. But feral cats have their source in the pet population, and there is hardly a farm on the island that does not keep a few barn cats to help keep rodents under control. We could not get rid of the domestic cats, so there would always be a source.'

There are around 12 million domestic cats in the United Kingdom, and an unknown number of feral and stray cats. The Mammal Society estimated that these cats kill an estimated 275 million wild animals a year, of which around 55 million are birds. And while cats may predominantly kill wild mammals throughout the year, the birds are easiest to kill during the breeding season when adults are incubating or juveniles are vulnerable.

These figures might seem large, but compared to the data from the US, where a 2013 study put the number of birds killed by cats at 2.4 billion every year, they are relatively modest.

It is argued that cats tend to kill the weak and sick who might have died anyway – and that might be true. But in that case those deaths would have gone on to feed other wildlife, rather than the already well-fed domestic cats. And yes, it is undoubtedly true that the change in the way our land is managed has had a dramatic impact on the ability of our wildlife to thrive, maybe putting the damage caused by cats into the shade.

But cats do more than just kill – they also disturb. With or without a bell (and the only bells I like are so large the cat has trouble lifting its head) a prowling cat will cause alarm. Parents will leave nests of eggs or young to try and distract the predator, in turn leaving the nest vulnerable to other opportunists.

I should be honest here, as I have tried to be throughout. I have a problem with cats. Yes, I love the fur, and many an hour I have wiled away stroking a purring bundle. But I have over the years been really enjoying the birds in my garden. Not like most people do – with feeders and flocks – no, I have been forming relationships with individuals. Robins, to be precise.

Since I wrote my second book, *The Beauty in the Beast*, I have been obsessed with taming robins to feed from my hand. Or rather, I have put myself into a position to be trained by robins to provide them with food from my hand. I have spent hours standing, holding my camera set at 1/4,000 of a second, on the patch of my patio that catches the morning sun, which I reflect back to the point where I hold my baited (dried mealworms) hand. And in those brief moments I have had some of the most beautiful wildlife encounters of my life – and this includes all the time out with hedgehogs, the sightings of whales, and the walking through Tanzanian, Namibian and South African wildlife parks.

So when the neighbourhood cats start prowling around my garden I resort to the pump-action water pistol usually reserved for summer water fights and trick-or-treaters. And when people use our (amazing) street WhatsApp group to start praising their feline companions, I pop up photos of 'my' robin.

Dealing with people's pets is not easy. The knowledge that my beloved rescue hound could cause damage to wildlife worries me, so he is always on a lead in sensitive areas. New Zealand has already broken ground on cats, but closer to home, in the south-west German town of Waldorf, a new scheme was introduced in 2022 that sees great restrictions being placed on pet cats.

For the breeding season of the locally rare crested lark, which breeds around the town, cats will have to be kept indoors. The bird nests on the ground and is very vulnerable to predation. Waldorf's other wild animals, like magpies and foxes, are being dealt with in a more forceful manner. The cats, well, it comes down to incentives for their humans to keep them locked up. Residents who fail to keep their cats under control from April to August (until 2025) will be fined €500. And if a cat kills a crested lark, that fine could rise to €50,000.

On Orkney, for now at least, it seems like cats are off the target list. But I do worry about whether success, which is measured in living birds and voles, not dead predators, can be achieved while there are still such effective killing machines running loose.

Every step of the way is recorded. They have a mapping app, which tracks where Spud goes and is used to collect the data from each trap. If an animal is present, the body is collected to be checked for sex, age and also for its DNA to be sampled. This time round they were all empty, but had the bait refreshed. The app also records the time, the weather, and makes sure the kettle is on when they get back to the central office.

Maybe not the last option yet, but this technological wizardry is all for a very good reason, as Sarah explained. 'We need to learn from what we do – that if we spot something is not working, we need to change it. And we also want to make sure that anyone else following in our footsteps can at least see what we did and use that as a starting point.'

The data they collect allows them to model the population of stoats. Now, this is when I started to do to Sarah exactly what journalists do to me when they want to write about hedgehogs. 'How many hedgehogs are there in the United Kingdom?' they will inevitably ask. And I will have to say I don't know. But what we do know is how the population had changed since we started systematic recording back in 2000, and the answer we have is not great. Between then and our 2022 report we showed that urban hedgehog numbers were down by 30 per cent and rural populations down between 35 and 75 per cent.

So it is with stoats; they do not know how many there are, but they have modelled the data collected from the traps and that gives a fairly good idea of what is happening. When an area is set with traps, over time the proportion of adults caught should drop. The un-trapped population will mature over a few years, allowing a number of different generations of adults to appear. If the traps are taking adults and juveniles at the same rate, then over time the percentage of adults caught should fall, allowing people cleverer than me to play with the numbers and give an indication of the total population fall.

When an area's stoat population has dropped by 90 per cent, the team move onto the next phase, which involves response traps – which is what we were doing with Spud: responding to a trap going off, searching the area, setting more traps if new scent or scat are found.

I looked at the clock – we were nearing the end of our allotted time. There was an important budget meeting at 3.00 p.m. and I wanted to give them a break – I had been asking questions since before 10.00 a.m. and it had been pretty non-stop. And talking of finishing – that was an important question – when did they expect the eradication to be complete?

'Well, we already know my original timeline is a little out of sync, thanks to Covid,' Sarah sighed. 'I had hoped that this would all be over by the start of 2025, but the reality is, it will take longer. Let's say, I know we won't still be doing this in 20 years.'

Lianne had been quite quiet during this part of the conversation, but at this, she leaned forward and said, 'This will end.'

'And while we are obviously going to have to keep going with the trapping', Sarah continued, 'we must really make sure we don't let the social side of the project slip. So far we have delivered over 300 external events, there are 148 trained volunteers who have contributed more than 12,000 hours to this project. We need people onside, we need them not to forget, we need them to let us know if they see a stoat. So for that we continue going to meetings, we were with the NFU last week, we are in all the schools. Did you see all the pictures?'

I had – the walls are covered in hundreds of A4 sheets of paper with the children's favourite animals delivered with varying degrees of accuracy.

'We also have them writing stories, and poems. The legacy of the project will be far more than just our islands being free of stoats. We will, I hope, have encouraged a deeper love of the nature we share these islands in a new generation.'

Actually there is another legacy of the work on Orkney, and that on Scilly and Lundy – that of lessons learned. For example, in 2023 the RSPB employed someone to tackle the problem of ferrets on Rathlin Island, off the north coast of Northern Ireland. Initially the problem on the island had been rats – again they were devastating the bird populations – but someone thought it a great idea to try and use ferrets to control the rats, and also the rabbits. I think by now you have probably seen where this is going .... So they are now using poison to eradicate the rats, and traps to remove the ferrets, and good community work to instil biosecurity into the lives of those who share the island with these amazing birds.

Raccoons

Most of my favourite writers are women. Each Christmas I gift myself a couple of non work-related books to read just for fun, and it is becoming one of my favourite parts of that season. 2021 I scored hits with *Drive Your Plow Over the Bones of the Dead* by Olga Tokarczuk and *The Animals In That Country* by Laura Jean McKay. Both brilliant, both quite potty in their own way.

My 2022 treats were Tanya Shadrick's powerful and honest *The Cure for Sleep* and then, just as reasonable holidaying was coming to an end, the latest from Amy Liptrot. Her first book, *The Outrun*, was a marvel. In her second, *The Instant*, I was delighted to find a reference to something relevant for me – she was talking about the wildlife she would see in Berlin, in particular the raccoons.

Raccoons have taken up residence in the parks of the city. There are records of small releases of raccoons in central Germany in 1927 and around Berlin in 1935. But the real surge of these furry and dextrous beasts was during the Second World War, when many escaped from fur farms around the city.

There are potential conservation problems. Raccoons are omnivorous, and while they are very happy to feast on human leavings – my Berlin-living brother has noted the bin-eruptions as a sure sign they have been around – they will also happily dine out on the eggs and nestlings of Berlin's birds.

They have made their home in the city, and have become so comfortable that stories of them joining commuters on the underground don't make headlines anymore. Though they do still make news sometimes – *Spiegel International* reported in 2008 the complication that they caused for one of the city's landmark hotels as they tried to work out what to do with the furry guest who had taken up residence in their garage. Named Alex by hotel staff, the raccoon had a moment of celebrity status as stories appeared in the press. But attempts to move him on were thwarted by the law, which said that only if the animal posed a danger should it be moved.

The German for raccoon is *waschbaer* – wash bear, after the way it delicately washes food. They are not quite so highly thought of in the United Kingdom. In 2010 the *Daily Star* carried the headline 'Nazi raccoons attack Britain' in reference to the 'vicious pests' being spotted in County Durham.

With incisive skills the paper told a story of Hitler's raccoons and how they had rampaged across Europe and were now terrorising residents of Blighty.

The terrorising came in the form of one being spotted utilising a garden bird feeder.

While the Berlin reaction to raccoons is rather relaxed, that is not the case on Haida Gwaii, British Columbia. These islands, formerly known as the Queen Charlotte Islands, are home to more than 1.5 million breeding seabirds, many of whom are burrow-nesting. This was not something that the Provincial Game Commission considered when it released raccoons onto the islands in the early 1940s as part of efforts to support the fur industry.

By 1992 raccoons had spread through the archipelago and onto many of the smaller islands where the burrow-nesting birds congregated. The internationally significant bird populations include around half of all the world's ancient murrelets, as well as the preposterously named rhinoceros auklet. Actually, not that preposterous – these relatives of the puffin have forgone the clown beak and evolved a horn-like extension to the top of their bill, a rhamphotheca, that gifts the bird its name.

Helgesen Island was found to have around 12 raccoons resident in 1993, while counts of the birds showed that approximately 30,000 burrow-nesting seabirds had been eliminated since 1986. Numbers of rhinoceros auklets were down by 79 per cent, ancient murrelets by 83 per cent, and Cassin's auklets by 95 per cent. Of course, there could have been many other factors causing this, but the evidence does rather point to the raccoons. Many of the burrows were found to have been dug out by them and fresh scat was found around the colonies, containing incriminating feathers. And over the same seven years, seabird populations on nearby raccoon-free islands increased or were at least stable.

There were also knock-on effects. The rare Peale's peregrine falcons are declining in number. They nest in Haida Gwaii and rely on ancient murrelets as a food source.

Control was not going to be enough; eradication had to be the goal. Though not everywhere, some things are just not possible, so the aim was to eradicate raccoons from the islands where they would do most damage.

The technique conservationists developed was to hunt at night, with a shotgun, from a small boat, spotlighting the shore. This was not a free-for-all; skilled hunters were needed to ensure that animals were killed humanely. It was also only used outside bird-breeding time, as then the raccoons were foraging along the strand line. This was not a fast process. It took three years to remove the raccoons from Helgesen Island for example.

Burrow-nesting birds are bothered by other invasive species too. Sitka black-tailed deer were introduced to Haida Gwaii in the late nineteenth century and seemed benign. Unfortunately they trample the ground around the birds' nest holes, and eat the protective vegetation. In fact the deer are at the heart of a multitude of concerns. It is all too easy for me with my animal-centred mind to forget that invasive species can cause enormous damage to vegetation as well.

On Haida Gwaii the problems are both ecological and cultural. The Haida nation has a particularly strong relationship with its cedar trees, as described in the 2005 report, *Haida Land Use Vision*: 'Cedar trees are important to many other living things great and small. They provide habitat for forest creatures, some of which are an important feature of Haida crests and histories ... When a Haida person goes for bark, a pole or a canoe, trees are approached with respect. Their spirits are hailed in a song and thanked with prayer. A bark gatherer takes care that the tree will go on living.'

The deer, untroubled by predators, strip out the understory, removing the cedar seedlings and much else of importance to the Haida – food plants like the blueberry and the medicinal devil's club.

So, along with the raccoons, the deer have also been targeted as part of a large-scale ecosystem restoration project. This was not foisted on the community by external conservation bodies, but the result of careful communication and learning. Starting in 2017, multiple methods were used to eradicate deer from a series of islands. Stations baited with maize, cedar and apples attracted the deer into the sights of hunters. This was then followed by dog-assisted ground-hunting, aerial shooting from helicopters, and shoreline hunting from boats.

Nearly 600 deer were killed in the first year. Nothing was wasted; meat, hides and hooves were handed over to the local communities.

And researchers who have been monitoring the ecosystems before, during and after the cull have found new growth of native plant species and positive impacts to the seabird habitat, as vegetation has been able to 'green up' more than at any time in the last 50 years. Increasing habitat complexity and cover gives the burrow-nesting birds a greater chance of success.

CHAPTER 7

Minkicide

Small islands are relatively – and I stress that word – straightforward to clear of predators. But what happens when there is an invasive predator with exceptional skills helping to cause ecosystem chaos throughout the entire country?

Of all the aliens to make Britain their home, the most beautiful and deadly has to be the American mink. These mustelids, relatives of stoats, weasels and otters, were first introduced in 1929, because their luscious coats had made them a favourite of the fur trade. In fact I am writing this book with my mother's mink coat hanging next to me. While clearing out the family home after her death, I found the coat in the back of the wardrobe. I had assumed it had long been sold, but as soon as I touched it I was transported back to my bedroom, my mother leaning over me to say goodnight before my parents went out for the evening. I could smell her perfume, not on the coat, but in my memory, and I could not stop stroking the rich fur.

There is no justification for the farming of these beautiful animals in the brutal conditions that were, and elsewhere still are, common. But there is also no denying the power the fur has over me. Thank goodness we now have a dog; it looks a lot less strange stroking fur when it is still on a living animal.

Almost as soon as mink arrived, escapees were noticed. However, it was not until 1956 that they were recorded breeding in the wild. With hindsight it feels remiss that this was not quickly identified as a problem. Mink are remarkable for more than their fur. They are brilliant predators, superbly adapted to their semi-aquatic riparian lifestyle. As such, they have a dramatic impact on the ability of the wildlife that share their riverside habitat to thrive.

Mink are a problem in many different habitats. And as a result they are being killed on any island they try and make their home. On the mainland, though, there is one species impacted particularly dramatically: the water vole. I have always had a soft spot for the water vole. In fact I have recently written a book about them. Seeing one doing what they do best – sitting on some vegetation in a clear chalk stream, munching away on more vegetation, is a nature highlight. I have had the privilege of spending time with bestselling author Kate Long as she introduced me to the community of voles living near her Shropshire home.

While they are around the size of a rat – a rodent with a far less effective PR agency – they are really quite different when you get to know them. In fact I like to think of them as mini beavers – even the field signs they leave, of felled vegetation, have the same distinctive 45-degree angled cut as that left after beaver harvesting, only on a smaller scale.

Rats have more prominent ears, pointier faces, and longer tails. Voles have furry paws and a softer face. But the most common way of distinguishing them is in their response to disturbance. There is something particularly delightful about the 'plop' a disturbed vole makes as it dives into the water. Rats can swim, of course, but not as well as water voles.

The water vole has to be sensitive because it is very well loved – Kate Long listed the species that would love to eat water voles and it is so extensive as to make you wonder why they ever go out at all. However, they have evolved strategies to cope with most of the threats. For land predators, such as foxes and stoats, they can leap into the streams by which most live, and swim for safety. For avian predators, there are the burrows they make into the riverbank. And for aquatic threats, like many-toothed pike, they kick up silt and dart to their burrows.

And then came the mink. An animal that can swim with amazing agility, but can also snake its way into the burrows – this is particularly true of the smaller female mink.

The mink began their assault on the nation's riverbanks as they became established – escapees from poorly managed fur farms were swollen by numbers released deliberately. There is some debate as to who was most responsible for this – fur farmers blamed animal rights activists, and there were probably a few well-intentioned but ecologically inappropriate releases. But others could feasibly have come from the fur farms themselves, realising that their time was up as the fur-trade dwindled and cashing in on their insurance as they let the hungry predators loose.

Water voles were already under considerable pressure. Around 90 per cent of UK wetland habitat has been destroyed in the last 100 years. Intensive agriculture was destroying their riverside homes: drainage to 'improve' the land; agrotoxin runoff into the water; canalisation of streams; arable crops up to the edge; removal of riparian vegetation; cows trampling down banks. All these factors were causing the water voles to wobble, but the proverbial final straw came with the mink. Water voles have the unenviable accolade of being the fastest collapsing mammal species in the United Kingdom. Between 1989 and 1998 the population fell by 90 per cent. At the time of writing, it is down some 98 per cent from peak vole.

While male mink will catch and kill voles, the real damage comes from a female setting up a territory, because the demands of baby mink are easily met by readily available water voles. And she will work through the vole territories, maybe taking one a night, which will soon create another gap along the riverbank.

If we want to see water voles back on our riverbanks, or plopping into the waterway with such furry aplomb, we need to remove the mink. This needs to be done in conjunction with a general improvement in water quality and the way the riparian habitat is managed, of course. But if we managed to recreate a voley idyll and left the mink, all we would be doing is feeding the sink hole.

By remove I mean kill, of course. There is a perpetual temptation to obfuscate, but honesty is the best answer, and what choice do we have? Returning mink to the misery of a cage in the hope of reigniting the fur trade is hardly a humane alternative.

So we need to kill mink to conserve water voles – that seems clear. In fact, I would suggest that this is one of the easiest stories in the book – we have an instinctive fondness for the water vole, and the mink is easy to demonise. After all, most people realise that the beloved 'Ratty' from *The Wind in the Willows* is, in fact, a water vole, and in that tale it was a bunch of mustelids who were causing a lot of the trouble.

I asked my friend Tom Moorhouse, whose writing has often featured water voles, who would be the best person to speak to about mink, and without hesitation he directed me to Professor Anthony Martin. An email of introduction started the process, and eventually I was able to find a time that suited us both – and also enabled me to use the drive over to his home near Ely as a stepping stone for one of the most eccentric evenings I have ever been a party to.

The road to Tony's house was typical of the fens, following the line of the clay to try and provide some stability above the porridge of peat, but still conveying the feeling of driving over a water bed. So focused was I on looking for the name of the house that I drove by the first time, but on return realised I had been using the wrong cue – the great stack of mink traps outside the garage should have been enough.

The garage door slowly opened as I pulled up and Tony emerged from tinkering on an engine. The garage was full of tools and, while he was apologising for the clutter, I was just looking on, realising that he was clearly a man of many skills.

Tony was dressed in green – moleskin trousers and similar colour fleece, and looked far more twinkly than his years should allow. We went through to the house, shedding shoes on the way, for coffee. Most people with a conservatory have it tacked onto a house, but this room was flooded with light and integral to the building. Far more than a conservatory, in fact, it was like the best viewing rooms at a Wildfowl and Wetlands Trust centre because his 'garden' was a lake and it was full of wildfowl.

'It's taken me over 30 years of experimentation to get this right,' he explained as he handed me a steaming mug. 'Made a mistake with the goosander though, they were malevolent. Killed my goldeneye just for fun. They had to go. I've been lucky to avoid avian influenza, so far. The worst problem I have had was botulism. I built this lake 25 years ago and in that time there has been an awful lot of bird poo, especially the swans – there are up to 500 of them at a time – and all that was making it shallower, which helped the botulism to cause havoc. So I had to hire in a digger – oh, I love big machines – this was a 25-tonne beast and it allowed me to remove over 3,000 tonnes of mud out of there. I've spread it on a field out back, going to plant sunflowers in it.'

I had thought that talking with Tony would be tricky because there was so much going on outside, but he is such a fascinating person and has squeezed more adventure into his one life than most people would into 20, and the expected hour or so quickly became nearly five, leaving me feeling rather inadequate.

Tony has been involved in many other projects, the most remarkable of which has to be the work on South Georgia, the island in the South Atlantic, 1,400km east of the Falklands. It is perhaps most famous for being part of the action in 1982, when it was temporarily occupied by Argentina.

'Oh, it is a special place,' Tony reminisced. 'A bleak beauty that certainly did not impress James Cook when he and HMS *Resolution* were trying to find the expected great southern continent in 1775.'

Cook's diaries include the words, 'It did not seem probable that any one would ever be benefited by the discovery.' He went on to say it was 'of less value than the smallest farmstead in England'. Though that did not stop him going ashore and claiming it in the name of King George. This short exploration did not improve his mood. 'The inner parts of the country were not less savage and horrible. The wild rocks raised their lofty summits till they were lost in the clouds, and the valleys lay covered with everlasting snow. Not a tree was to be seen, nor a shrub even big enough to make a toothpick.'

But his observation of 'sea bears' – what we would now call fur seals – did spark interest. And within short order, millions of them, and later elephant seals, were hunted, the latter for oil. The slaughter extended, inevitably, to the whales that fed in great number on the plankton-rich waters.

With similar inevitability, the ships bringing death to the sealife brought with them rodents. The black rats never persisted, but brown rats and house mice made South Georgia their home, using burrows dug by seabirds. While Cook might have seen an inhospitable place, the rodents found plenty of food: an abundant supply of eggs, chicks and birds allowed them to flourish.

'The trouble with rodents on an island like South Georgia,' Tony said, 'is that while the rats and mice can have many offspring many times a year, the birds often lay but one egg per season – and many of the birds do not breed for the first few years of life.

'Entire colonies of birds were destroyed. The smaller species, storm-petrels, prions, and diving petrels, were effectively wiped out on the main island, retreating to the small offshore islands that were rodent-free. And we almost lost the only songbird on the island, the South Georgia pipit. That survived by clinging to a narrow strip of frigid land where even the rats could not survive.'

It is somewhat ironic that as the industrial destruction of seals and whales abated, they began a slow and steady recovery, whereas the wildlife that was not targeted for slaughter continued to disappear, thanks to the hitchhikers left behind.

'For a long time it was thought that this was an insoluble problem,' Tony said. 'New Zealand, as I am sure you are aware, have been leaders in the field of predator removal. But these have been from smaller islands. South Georgia is big – 165 kilometres long, 35 kilometres wide – and it is surrounded by some of the fiercest seas on the planet.'

Up until South Georgia, all similar projects had hit an entire island in one go: if you eradicate only a portion of the island, then that area can always be repopulated while you move onto the next. But South Georgia presented conservationists with one spark of kindness in an otherwise quite unforgiving landscape. Glaciers – rivers of ice running from the uplands – created barriers that prevented even the most daring of rodents from recolonising, as long as no more came by boat, of course.

Now, not all glaciers were barriers, as there were smooth ones that could be crossed by a determined rat. What the conservationists were excited by were the deeply crevassed glaciers that were utterly impassable. These allowed them to treat South Georgia as an 'island of islands', where each 'baiting zone' could be tackled independently. There were 19 rat and two mouse zones, ranging in size from just $4km^2$ to $233km^2$. In all there was over $1{,}000km^2$ to be treated.

This was not a simple matter though. Clearly none of this was simple, but there was an additional element of concern: climate change. The fact that some glaciers were retreating at a staggering 3m a day when Tony was there added urgency, because it would not be long before some of these very convenient walls of ice would retreat enough to allow rodents to recolonise what were previously islands.

So, in theory, this could work. If there was enough money for enough helicopters and fuel and people and poison and … the list is quite long. But really, could it work?

I am bypassing the process by which a small charity based in Scotland – the South Georgia Heritage Trust (SGHT)

– working with the Friends of South Georgia, managed to raise millions of pounds to make this happen. The details, along with much more, are written up in a lavishly illustrated book Tony wrote, called *Reclaiming South Georgia*.

The 2007 decision to tackle this problem led to a feasibility trip in 2010, and in 2011 the crucial trial was undertaken. This was to measure two important, and deal-breaking, issues. Was it possible to get helicopters to deliver poison in such a way that the rodents died and ensure it was all of them? And could this be done with minimal impact to the non-target species, mainly the birds?

The fact that I am writing about this means it should be no surprise that the answers were yes and yes. Which left Tony and his team with the daunting task of making it happen. The statistics are impressive. More than 1,000 hours of helicopter time, flying the equivalent of three complete trips around the world, dropping more than 300 tonnes of bait. All made possible by 39 people who spent nearly 4,000 days in the field consuming 10,700 meals and using around 10,000 teabags. The weather meant that even when the team was on the island, they could only fly on one day in six. This was a monumental undertaking, and was only possible thanks to the sheer will of Tony and his team. This was not work for the half-hearted.

I can completely understand the desire to move the rodents from the island, but what about the consequences of spreading 302 tonnes of bait across the landscape? Brodifacoum is very effective at killing rats, but what about the other wildlife?

'Two things,' said Tony. 'Firstly, we know that if we do nothing then there will be no potential unintended poisonings – but then we also know that if we do nothing the birds will be gone from this island. We can't trap the rats out of existence – can you imagine the number of traps and people that would be required to cover this much ground? Worse, thousands of those traps would have to be placed on vertical sea cliffs. And this is not a project where a 99.99 per cent reduction can be considered a success – it would just be a waste of time as the problem would reassert itself and even more animals would need to be killed. Secondly, this is formulated to discourage birds eating it. Apart from the pipit, most of the birds here are predators, so cereal bait is not attractive, nor is the colour – the bait was dyed blue/green. The trial season of 2011 allowed us to evaluate the impact of baiting on non-target species over a relatively small area and then either modify the methodology for future seasons or pull the plug entirely if that impact was unsustainable. That was a unique luxury of the South Georgia operation.'

The consequences of not acting are frequently overlooked. When I started hearing these stories my heart would lean into inaction, but my head would point out the reality. Our actions have created a problem. We can leave that problem, but not acting is still an active choice, and it has consequences. Similarly, acting half-heartedly – or acting only to be perfect – loses the good. Yes, it might be better if every invasive predator was given a dose of calming poison while being lulled with their favourite music, like in the thanatorium in the film *Soylent Green* (plot spoiler, it is people!) But that is, obviously, absurd.

The reason for introducing you to this story is to explain why I think what Tony has in mind now is not as ridiculous as it might at first seem.

With coffee, sat by the large windows of his observatory/conservatory, we were about to dive into the subject at hand, when Tony decided that we would also benefit from the wood burner. It was a grey day. While he got it loaded and started with practised ease, he asked me about hedgehogs and the Uists. He was not shocked that the problem remained unsolved. I get the feeling that he has limited patience with committees. He likes to get on and do things.

So, mink. We know we can do all the good land management we want to aid water voles, but if their worst enemies are still in place, there is no hope. And locally, we can remove mink and see benefits within a few years. For example, survey work was carried out on the River Deben in Suffolk in 1997 by the Wildlife Trust. Along a 15km stretch, 80 per cent of sites were occupied by water voles. By 2003 the area was affected by mink and a repeat survey showed this figure had dropped to less than 50 per cent. Over a three-year period, 101 mink were killed on this stretch, in parallel with habitat enhancement, and the next survey showed water voles back and occupying over 80 per cent of sites again.

But local projects are really only tiny plasters on a serious haemorrhage of life. Each patch has the opportunity for re-invasion if there are still mink in the wider landscape. And if we were to start a nationwide campaign to remove mink, it is not going to be 'done' in just one year, so there will always be the chance of their recovery. It might seem strange to look at the rigorous South Atlantic island with fondness, but South Georgia did at least have barriers to prevent the return of rodents. We do not have that here.

'I learned to skin mink in my youth,' Tony said. 'It is a strange circle of life to have ended up with them again. You see, my father used to run a small mink farm in our garden on the Sussex coast, to supplement his income as a bank clerk. So I was around them from an early age. And it feels good to be trying to undo some of the damage unwittingly brought about by fur farming. In fact, more than some of the damage – I believe we can eradicate mink from mainland Britain.'

I paused, coffee midway to mouth. I knew Tony was ambitious, but I had no idea he had set his sights quite so high. Mink control has been part of the work of conservation groups for decades. But eradication – I had thought that was just a pipe dream.

'As I am sure you have found out, this sort of work can only succeed with public consent, and support,' he said. 'I recognise that there are some people who don't want to see any animal killed, and I sympathise with that view. But I know – we know – that so many more animals are going to die if mink are just controlled. Both mink and the native animals on which they prey. It is a question of whether we are willing to take responsibility for the problems caused by our ancestors, quite literally in my case.

'Membership organisations are sensitive to discussing this sort of work,' Tony continued. 'They fear a backlash. But I believe if we are honest, we can win most of the people over. A signal moment for me was when some hunt saboteurs posted on their Facebook page photographs of mink traps they had vandalised, crowing about their destructive prowess. I was astonished, but pleased, to see that the vast majority of the responses, all from animal rights supporters, were critical of the damage. I do wonder if there may be some residual collective conscience at play here – a recalling of the fact that thousands of farmed mink were released into the wild by "animal liberationists" back in the 1970s. They of course had no clue as to the appalling damage that this well-meaning action would cause.'

I wanted to get back to his dream, and wondered why he was aiming at mainland Britain when there are already eradication projects on our surrounding islands.

'Islands like the Outer Hebrides might seem a sensible place to start eradicating mink,' he said. 'But such wild, remote islands present huge challenges. Toxic bait can't be used, as we did with rodents on South Georgia, and has been done in many parts of the World, because native species would also be poisoned. So trapping is the only option, but live traps can only be used when you have people to check them frequently, or if they are equipped with an electronic box that tells you when the trap door closes, and even then trap access has to be guaranteed. A mink must not be left to starve in a trap because a storm meant a boat could not get to an island for a few days. The humanity with which this work is conducted is as important as the efficacy.

'These electronic boxes [Remote Monitoring Devices – RMDs] are game-changers, in that they dramatically reduce the number of times a trap must be visited. They tell you, with absolute reliability, when a particular trap door has closed. You receive a text message and email within 90 seconds. Without an RMD, the daily visits quickly become boring, especially when mink are few and far between. With one, a routine visit is only needed every three months, and volunteers remain with us, happy to keep their traps active and vigilant year-round. High-tech meets low-tech in a very productive way.'

With this idea of eradication in mind the first job, four years ago, was to see what others were doing in East Anglia. He is not completely mad, he was starting relatively small. There has been a long-running Norfolk Mink Project, which has always run on a shoestring by Simon Baker, a quiet, retired, visionary civil servant who co-led the campaign to eradicate coypu from England in the 1970s . Suffolk and Essex Wildlife Trusts were doing their best to stem the mink tide, but never with sufficient resources to do more than this. Cambridgeshire and Norfolk Wildlife Trusts had some low-level trapping work, but never enough to make a landscape-scale difference, and no one had an idea of how many mink there were out there.

'Simon and I developed a plan to see if it might be possible to do what most others had dismissed as impossible - eradicating American mink once and for all, rather than catching them on and off in perpetuity. If so, it would be cheaper, quicker, more humane and far more beneficial for native wildlife than the status quo. For this we'd need to encourage the entire, diverse mink-catching community to work together towards a common goal,' said Tony. At this moment we were interrupted by the sound of a hunting horn. It was an alert on Tony's phone. 'That will probably be a water vole in a mink trap,' he said, popping the phone back into his pocket.

I wondered whether he needed to do something, like go and rescue it. Or, if it was actually a mink, kill it. 'I can't do East Anglia alone,' he said with amusement. 'A whole team of people just got pinged, the nearest will deal with it.' I was only beginning to realise quite what he was up to with this project.

'We decided to give it a go for two years and see if it could work. I am not getting any younger, and while I want a good project to sink my teeth into, I had no intention of wasting my time and energy if it really did turn out to be an impossible task.'

They raised some money – the Heritage Lottery Fund came up with £230,000 – and were able to invest. The traps I had seen outside the house were part of this, but traps alone are not enough. Over the years the art and science of minkicide have been developing. In 2002 the Game and Wildlife Conservation Trust developed the mink raft, as a way of both detecting and trapping mink.

Detection was by means of a layer of wet clay on the floor of the raft that could be left for a week or two, on which the footprints of any inquisitive visitors would be recorded. If and when mink prints were found, then a trap could be set, and if mink were still in the area, they should be caught within a few days. This strategy worked well in reducing mink numbers, but was labour-intensive and led to many mink passing through without being caught. The introduction of RMDs meant that traps could be left on the rafts permanently, and that any visiting mink was detected not by its footprints, but by a pair of beady eyes greeting the volunteer trapper when he or she was summoned by a text message on their phone.

'I hate killing animals,' Tony stressed. 'And if there was another way to restore our riparian ecosystems without killing mink, then I would be all for it. But until then, it is essential that we do it humanely.'

When a mink is caught – or anything for that matter – and the alert has been sounded, one of the team of (mostly) volunteers heads to the triggered trap. If it is a water vole, water rail or other 'good guy' it is gently released. If it is a mink or grey squirrel, neither of which can legally be released alive, the animal is dispatched with a single pellet from an air rifle.

The bodies are collected and stored because they help tell the story of how the population is doing. 'The teeth, for example, allow me to age the animals', Tony said, 'which is crucial in modelling how the population is changing. Lots of young mean they are still breeding, only finding old ones suggests they are not. Eventually they are recycled. I have some very content buzzards, magpies, crows, jackdaws, mice and, yes, hedgehogs at my place.' I did wonder whether he was tempted to keep the skins. 'I did keep the first 100 that we killed; I imagined I would have time to make a coat and if the project had failed, maybe I would have!'

This is all good and honest but I still struggle to see how this can be scaled up to cover the county, let alone the country.

'Like I said, I am not doing this if it is not working, and the results have been amazing,' said Tony. 'People care about water voles and more importantly, they are willing to act. So, in East Anglia, we have around 700 volunteers. We store and maintain around 800 trapping rafts. And these get distributed to every river, stream, drainage ditch and lake. We also put them out in response to mink sightings.'

The reach is more than just these volunteers of course – education is key. 'I want people to be aware of mink, of what they do. I also want them to be aware of what we do, and perhaps more importantly of what would happen if we did not do what we do. But I don't want to demonise this animal. I can't stress enough, these are glorious creatures, it is just that they are the wrong animal in the wrong place at the wrong time. They are not the devil. We humans made a mistake by introducing them to this country, and it's a mistake which we can and should rectify.'

These sorts of projects always take an age to show signs of success, you might think. Especially when spread over such a wide geographic area. When Tony told me what had been happening, I was absolutely taken aback.

'During lockdown we had to rely on people in their own patches keeping on top of things, but as we opened up again, well, we were able to increase the pressure. By the start of 2022 we were still catching quite large numbers, but they were down a third on 2021. Then suddenly the catches fell to zero.'

Now that might not be good news; maybe mink had learnt about traps? Maybe there was a better food source?

'We increased the number of traps,' Tony said. 'Made sure each was fully charged up with the scent lure, expecting August – when the young start to disperse – to see a surge. But still, almost nothing. Just a few from the west of Cambridgeshire, where reinfection would be easier, and from Essex where there had been less trapping. And it was not just the traps – members of the public were not seeing them either. The population had simply dropped off a cliff.'

So many questions were bubbling up for me but before I could launch into them, Tony answered one.

'The scent we use is eau de mink,' he said, getting up. 'Come along, I think you will like this.' I followed him out to one of a number of freezers in his garage. 'One of the things I do with the mink that come in from the cull is to harvest some of the scent from their anal glands. I have these syringes, and with a gentle squeeze of the glands, with thumb and forefinger, either side of the anus, I can collect a drop or two onto each of maybe 20 cigarette filters. And then I store them in here, before using them on my traps or posting them around the country.' He put on a pair of gloves, and offered me a pair, and reached into the freezer, taking out a small packet of filters, about 15mm long. 'These are the perfect shape and size. I take one and pop it in a plastic practice golf ball, you know, the ones with lots of holes, and suspend it above the treadle in a trap. The lure will last for three months. Here, have a sniff,' he said as he handed me one.

Oh my goodness. What a smell – like onions forgotten in the bottom of the vegetable rack that were then stored in the armpits of someone who had been sweating nervously for a week. Tony took pleasure, I think, from my discomfort. 'Remember, these are frozen, this is just a fraction of their full potential. Imagine what it is like when they are fresh and all those volatiles are swarming on a summer's day up your nose. And I reckon it stood me in good stead while Covid was rife – no one would come within 20m of me.'

Clearly these lures work. 'If I were a betting man I would say that there were up to five breeding females in all of Norfolk, Suffolk and Cambridgeshire. That is 10 per cent of England.'

A very small team with a relatively small budget has transformed the mink-map of Britain. For the first time, I sensed a potential end of mink, and more importantly, the return of the water vole to our rivers. What was not clear, though, was how this happened. Tony has a theory.

'I have been surprised. I imagined there would be a drop and then a plateau – a long difficult journey to try and find the last few. But I think something else is at play, and that is the "Allee" effect.'

I had to admit to ignorance; my student days were a long time ago. And it is a fascinating bit of counter-intuitive population ecology. You might imagine that at a higher density, a population would suffer from increased competition for resources and have a reduced growth rate, and that at a lower density the population would be able to increase the growth rate due to less competition. And that is true with many species. But some species display a reverse of this effect, one that got its name from Warder Clyde Allee who first described it back in the 1930s.

'The best way to look at it is as the opposite of overcrowding … so, "undercrowding"! When the population gets below a certain level something happens. For some species, it is if they hunt together, but with mink, I think it is simply down to males and females not meeting. Remember, these animals can easily cover a five-kilometre territory of riverbank. We have opened up every female we have caught, and there is no evidence of breeding.'

My next surprise was down to the fact that I had not heard anything about this, as surely this should have been shouted from the rooftops.

'Despite the success, despite the almost unanimous support for removing mink, I find the Wildlife Trusts to be hugely hesitant about publicising this work.'

Membership organisations are in a bind: relying on the support of members of the public, they fear rocking the boat. Maybe it is my turn to be naïve, but surely if organisations are honest, explain what they are doing and why they are doing it, members will understand?

But even as I write that I know the twitching trigger-fingers of the underemployed outrage warriors are getting ready to launch into a Twitter pile-on.

Social media, though, is not all about knocking the work of Tony and others like him. 'Facebook has proved to be amazingly useful; the Norfolk Wildlife Trust has around 5,800 members on the site and many of them have cameras. They post their images, some of which are, or were, of mink. The eyes and ears of the public are vital, and they have shown how quickly the water voles have rebounded as well.'

The great news is that this is not expensive kit. Well, it mounts up, but compared to paying a team of full-time operators or running a campaign on South Georgia, it is doable. The rafts cost £115, the traps a further £50 and then £120 for the remote monitoring device. With a few other bits and bobs and delivery you are looking at £330 for a smart mink raft.

Tony had used the opportunity of going to the garage to load up a bucket with bird food and I followed him out into the 'garden', where the tamer fowl recognised this moment and came dashing our way. Spending time with nature around – the sights and sounds of these birds – it is a wonder that he ever gets any work done. Worse still, he pointed out his office in a small turret-like affair, overlooking the water. This is a greater measure of his willpower than the work on South Georgia, that he can pull his gaze from all of this.

'It's not just wildfowl here,' he said, as he tossed another cupful of food into the water for the hungry smew – tiny, pretty birds. 'You know you have talked a lot about hedgehogs, and how their numbers are declining. Well, I think I have worked out why .... They are all here!'

The reason the hedgehogs like it so much is that of his 30ha of land, 25ha are enclosed in badger/fox-proof fencing. And while 5ha are given over to water, there is plenty of space for other wildlife to thrive. 'You see that concrete block,' he said, pointing to beyond the pond. 'It is full of holes. I built it for sand martins. In the summer I have had up to 44 pairs of avocet, little ringed plovers, 2,000 pairs of black-headed gulls, Mediterranean gulls too. And winter – well, you see here, the geese, swans, teal, gadwall.'

Heading back in, he regains a more reflective mood. 'I think it is now logistically feasible to eradicate mink from mainland Britain. These smart traps – and all thanks to Jonathan Reynolds at the Game and Wildlife Conservation Trust for designing them – mean we can be humane and free of non-target mortality and negative environmental impacts. We know how much British biodiversity and bioabundance would benefit – it is not just the water voles who suffer. I think, given the resources, we could do this in a decade, though this would be dependent on co-ordination between many conservation, fishing, farming, and water-related organisations, together with the consent of landowners.'

A mink-free Great Britain would plausibly cost tens of millions of pounds, but this has to be weighed against the limitless future costs of mink control. It would be by far the world's largest invasive predator eradication project by geographical area and would set a precedent for citizen-led conservation action globally.

Tony has gentle confidence, usefully coupled with self-awareness. 'I really believe this could work. I just need to find an excellent, small team to take on the bits I am not good at – I can be a bit like a bull in a china shop when I feel we are moving too slowly. And we cannot afford to move at the pace of the slowest. We can't wait for large organisations to have more meetings about the wording on yet another report.'

There is often a critique of this sort of ambitious work – that it is playing god. This is something that Tony has considered a lot. 'There is an ethical and moral dilemma about whether it's acceptable to kill a "bad" animal (a rat, a mouse or a mink, say) in order to save a "good" animal. The debate is made even more complex when, as is the case with mink in Britain, the bad guys doing the damage were introduced by humans. Is it not right that subsequent generations of humans, quite literally in my case, should repair that damage if we can? I don't claim to be right, and others to be wrong. But I can't stand by and watch our precious native wildlife disappear if I can do something to prevent that.'

In 2020, Tony cowrote a paper for the journal *Mammal Review* in which they set out the sequence of six questions that have to be answered before a conservation project can take place, and how in each case, if the answer is in the negative, that is enough to shut down the entire process. They are a valuable guide that could and should be applied, with situational variation, to all such endeavours.

1. **Can all individuals be put at risk by the eradication technique?**
   The vast majority of mink will be trappable. The final few per cent will prove to be extremely challenging, especially in remote locations. Those on the uninhabited islands and coasts of north-west Scotland would take substantial resources, but in principle all mink could be put at risk.
2. **Can they be killed at a rate exceeding their rate of increase?**
   Experience already shows that for mink, the answer is yes.
3. **Is the probability of pest re-establishment manageable to near-zero?**
   Possible means of re-establishment are by:
   (1) vehicle on a ferry or train within the England–France Channel Tunnel
   (2) walking through the Channel Tunnel itself
   (3) escaping from zoos or private collections
   (4) deliberate reintroduction
   That several large islands served by ferry are still mink-free indicates that (1) is extremely unlikely. The risk of (2) is extremely low because of the length of, and environment within, the Tunnel. Zoo escapes are possible, and measures would be needed to reduce that risk to near-zero. Deliberate reintroduction is a risk, but a small number of animals is unlikely to become established if they disperse, and constant vigilance ought to detect a new population before it can spread widely, allowing the animals to be found and destroyed.
4. **Is the project socially acceptable to the community involved?**
   Although there may be objections, mink control in Britain has been carried out for decades with relatively little resistance, perhaps because the wildlife benefits of the work are accepted. Any landowners determined to

protect mink could present problems, although statutory measures are available to deal with this. In their thoughtful review of the water vole/mink dilemma in Britain, Tom Moorhouse concludes that 'there is no shortage of public support...for mink eradication'.

5. **Do the benefits of the project outweigh the costs?** This is a subjective judgment, but millions of pounds are annually committed to eradicating invasive alien species in Britain, including mink, so the concept is not new. In financial terms, the cost of an eradication would, in time, be less than the ongoing costs of control and mink damage, so the logical answer to the question is yes.
6. **Can animals be detected at low densities?** Yes, mink leave visual and olfactory cues of their presence. Trained dogs are readily able to detect single mink. New molecular techniques should soon be available to simplify the search for mink at low densities and render the task of finding them much less expensive.

Tony Martin is a fascinating mix of characteristics. He has a rigorous academic intellect; that is obvious. He has a can-do practical skill set that would embarrass most engineers. And I also noticed that in conversation he would often personalise, but not anthropomorphise, animals. They get a who, not a which. This is something I found myself doing more often. I think it is important in the way it encourages us to think about the non-human animals we share this planet with. 'It is not complicated; they are individuals,' he said. 'Every one of the sand martins here has been ringed, the Amazon river dolphins I have the privilege to study, they are all known individuals. We can identify the whales that we encounter and learn their families. These are sentient animals. They deserve our respect.'

I was very conscious that I had taken up a lot of Tony's time, but he was still very keen to talk. As I took my coffee mug back into the kitchen I noticed a case of wine had just been delivered from the Wine Society – another shared love. I could think of little better than a summer evening watching the birds, sipping wine, in his company.

'There is a lovely story,' he continued. 'A farmer down in north-west Essex, on the River Pant, told me how he remembered seeing water voles as a child. But never as an adult, so he would trap mink. He described it like catching mackerel at first, and then one day the trap went off and when he went to look, there was a water vole. First he had seen in 50 years. Crusty old farmer utterly blown away by this, and all thanks to getting rid of mink.'

And now, almost every single cry of the hunting horn is a water vole – amazing that the well-trodden aphorism, 'Think globally, act locally,' should still have such resonance.

'The great thing is that if we can eradicate mink, the chances of their return are small.' But there is always the unexpected. Tony gave me a sobering example, from one of the most high-profile attempts to control rodents. Gough Island makes South Georgia seem quite cosmopolitan. A dot in the South Atlantic, it is part of the British overseas territory of Saint Helena, Ascension and Tristan da Cunha. Just 13km by 7km, what it lacks in area it more than makes up for in ornithological importance. It is home to one of the most important seabird colonies in the world.

The RSPB reports that there are more than 8 million breeding birds from at least 24 different species using Gough. These include highly threatened species such as the endangered Atlantic yellow-nosed albatross, Atlantic petrel and MacGillivray's prion, and the critically endangered Gough bunting and Tristan albatross.

By now it won't come as a surprise to learn that mice have made their home here, thanks to nineteenth-century sailors. But these are no ordinary mice. Over the years they have learned to exploit the rich food resources available at breeding times. And they have been videoed doing so in most graphic detail.

The mice feed on the flesh of live birds. It was thought to be just chicks, but now adults have been seen on camera being eaten. Oh, and these are not small birds – the chicks of the Tristan albatross weight up to 10kg and the adults have a wingspan of up to 3m. The gnawing mice don't, obviously, consume entire birds, but the open wounds frequently lead to the death of chicks. The mice have transformed, and are now weighing in at around twice the weight of your average mouse.

This might seem absurd. Why don't the birds defend themselves? We are prone to thinking anthropogenically – the human lens is clear. But that is not the view of the world from an albatross eye. These magnificent birds have never encountered a ground predator before, they have evolved no defence. Their reaction to threat is to stay still. Even when that threat is eating them alive.

The RSPB reported that 'the situation was so severe that just 21% of Tristan albatross chicks survived to fledge during the 2017/18 breeding season. And for endangered burrow-nesting species breeding success has been recorded as low as 0%'.

So they embarked on an ambitious programme to try and save the birds on Gough, by eradicating the mice. This was undertaken in 2021. The RSPB team was well aware of the scale of the task, noting that 'the solution was relatively straightforward ... though the operation to do this was logistically complex, mainly because of the island's remoteness, tough terrain, and harsh weather conditions'.

The nearest city was Cape Town, some 2,700km to the east. The nearest point in South America is 3,200km away, so Gough is properly remote. They employed the same tactic as on South Georgia, just 2,400km to the south-west, of helicopters and poisoned cereal bait, spread during 2021. They took the precaution of removing safeguard populations of land birds so they could be kept out of the way until the operation was finished.

Unlike South Georgia, after the eradication attempt had been completed, mice were found to still be present in 2022. Now, this is not the end of the world, as the reduction in mice meant that there was some respite – the breeding success of Atlantic petrels, for example, more than doubled.

The team are not heading back to repeat the 2021 treatment, at least not immediately. They are trying to learn why that attempt failed, as there is no point doing the same thing again and expecting a different outcome. That way lies madness.

'One single lesson I got from the South Georgia work', Tony said, 'is that with a good plan, good organisation and a can-do attitude, a tiny organisation can achieve wonders. Prior to the rodent campaign, the SGHT had undertaken modest, tourist-related projects. But just ten Trustees, with no experience of anything like this, then had the vision to initiate the world's largest rodent eradication. Though to be honest, I think if they had fully appreciated what a monumental challenge was involved, they would never have started what, by any standards, was a barking mad idea!'

Getting the right group of people together is something of an art. 'We did South Georgia with a minimum of bureaucracy, a clear goal, and an absence of ego. This is about the work, not ourselves. Oh, and a can-do attitude from a wide range of disciplines. If you can get the fundamentals right, you can achieve miracles.'

If Tony could find a way of guaranteeing these qualities in other walks of life, just think what could happen!

'You ought to have a look in here before you go,' Tony said as he walked me towards a large chest freezer back in his garage. On opening it the first thing I noticed was a tail. 'Fox,' he said, picking it up. 'I couldn't just leave it, look how gorgeous it is.' He handed it to me. It was quite amazing, though we both agreed they tend to look better on their host. He was rummaging around for a bit longer and pulled out a piece of blond fur, again handing it to me.

'This is one of the last pelts from my father's farm,' he said. 'See how blond this is – a recessive trait. Most of the mink were pale on the farm. And as long as two blond recessives are mated, then you get more blond. Then in the sixties there were jet black mink brought into the mix. Same again, recessive traits – normally they are brown. And within a very few generations most of the mink in the wild had reverted to that colour.'

Next out of the freezer was a piece of flattened fur, and it was clearly a prize possession. 'This is the skin of the very last rat on South Georgia.'

I left Tony feeling more hopeful than I had in a long time for the future of our water voles – and headed to, as I have already alluded, a very unusual night. A few years ago I set up a petition on change.org, calling for hedgehog highways to be incorporated into all new housing developments. It sparked attention and at the time of writing has well over a million signatures. Part of the motivation for this campaign is the mailing list of all signatories, who get sent updates every couple of weeks or so. I try not to bombard them with too many hedgehog stories! But communication has always been tricky around the petition, conversations hard to manage. So I set up a Facebook group – hedgehoghighways – and recruited two amazing moderators to keep the tone civil.

Okay, that is the background. The husband of one of the moderators, Rose Sewell, found out that we had never actually met, despite working together for years. So he asked whether I would be willing to be her birthday present, and to surprise her by turning up for a dinner out. And that is what I did after my day with Tony Martin, because I was in roughly the same part of the country – it was such a wonderful opportunity to make Rose happy. First time I have ever been a birthday present!

Pythons

I am torn, as I drive, between taking the time to listen to music or books. Music is on in the background at home, while I am cooking and cleaning. But to really listen – that is a luxury often only available in the concert hall. Or when I make a conscious effort in the car. The last long journey to my mother gave me space to listen to Bach's *St Matthew's Passion* in detail – so beautiful.

But then there are the books. There are so many that I know I should read or at least listen to, but somehow self-improvement is harder to countenance on the motorway. Looking for something light in spirit, but substantial enough to sustain, is the trick. The last time I visited my old friends Gillie and Dave I was given a recommendation that was to prove enormously fruitful. Carl Hiaasen has long been a favourite author, though as I rarely read fiction these days I had been out of the loop. Well, his 2020 outing with *Squeeze Me* was perfect.

Hiaasen's work is laugh-out-loud funny, with a social justice and environmental message not so subtly hidden within these Florida-based stories. What makes them even more amazing is that the absurd situations he describes are taken, in part, from real-life events. Florida has many strange characters, in Hiaasen's work and in real life. *Squeeze Me* features a very Trump-like president and his Florida 'White House', which doubles as a club for hire. It is full of grotesque characters and the odd hero.

The point, for this book, is that while I was driving and listening and being amused enough to relax into the sluggish curl around Birmingham, the heart of the story was revealed; Burmese pythons. A delightfully mischievous character causes chaos by releasing magnificent examples of these non-native snakes in and around the lives of the obscenely wealthy.

Burmese pythons came to the US as part of the exotic pet trade. There are records of them back in the 1930s but the pet-keeping really took off in the 1980s. They are easy to breed in captivity, as females can lay up to 80 eggs in a clutch. The hatchlings are quite large – up to 60cm long – and are already able to eat adult mice, making them easy to feed. They are beautiful.

But very few pet keepers, who are excited by a small youngster, will be equipped to look after an adult that can live for 20 years and easily reach over 5m in length. As the snakes grow, they get discarded by their owners, and many have made their homes in the Florida Everglades, where they were first sighted in the 1990s.

It is argued that a hurricane in 1992 might be even more important in this story than irresponsible herpetophiles. When the storm hit, apparently more than 700 pythons escaped from just one of the many import facilities. Their import to the US was banned in 2012, but that was very much a case of shutting the door after the snake had bolted. Estimates for the Florida population of these reptiles is now in excess of a million.

They are now classified as an invasive species and have a dramatic impact on native wildlife. Studying roadkill is a useful way of seeing what is living in an environment, and the comparison over time can reveal how things are changing. So when the 1997 data is compared to the 2011 findings – that is before and after python proliferation – the declines in raccoon, opossum, bobcat, rabbit, fox and other mammals ranged from 88 to 100 per cent.

This could have been coincidence. But the species suffering the most dramatic declines were also the ones turning up in the stomachs of dead snakes. In the end, it was an alligator that really got the snakes into the news, as in 2005 a photo appeared of a dead python with a 2m alligator that had exploded from its stomach. It is not always that way around, and in 2023 a viral python/gator video of the moment featured the alligator in ascendence.

There is good news, in that the ability of these snakes to further invade the US is going to be limited by climate. They need the swampy warmth of Florida to thrive. Though that was only good news until the US Geological Survey reviewed how climate change might alter this; the future is great for the snakes, less so for the country. They predicted that by the end of the twenty-first century these pythons could have made their home in a third of the continental US. Though to be honest, if conditions change that dramatically, the snakes are going to be the least of people's worries.

Trying to control this species is particularly tricky. The Everglades has large areas that are pretty inaccessible to people. Dogs are being used, but this could only ever work around the easily accessible areas. Traps fail due to the snakes' enormous ranges (and a tendency to eat and then not move for a long while), though they could be used as a defence around critical areas. Biological control gets talked about as a potential way forward. But any introduced disease would have to be absolutely species-specific to avoid disastrous spillover into native species. Others have suggested reintroducing jaguars into the Everglades, as these big cats hunt anacondas in South America. But I can't see that being embraced by the tourism industry.

Another approach, not from the science community, is the use of bounty hunters, which has received a great deal of public support. But the results are not exciting. The 2013 Python Challenge – a month-long event with cash incentives, sponsored by the Florida Fish and Wildlife Conservation Commission – resulted in only 68 captures by 1,600 participants. A repeat in 2016 scored 106 snakes from more than 1,000 entrants. This is not easy work. While some people have tried marketing snakeskin products as part of a conservation exercise, eating the pythons is not recommended due to the high levels of mercury that have bioaccumulated.

There is still no plan on how to control the pythons. Unchecked, they will massively change the ecosystem of this part of the world.

All of this just added to Hiaasen's fancy, and the eccentric renegade who unleashes pythons around the presidential party – while tending his reptilian friends with care – accepts their inevitable doom at the hands of the security services, as they are in the wrong place at the wrong time. It is not the snakes' fault that they are causing chaos in the Everglades.

CHAPTER 8

Keepers

As will already be apparent, I really struggle with the idea of gamekeepers; people who kill wildlife for a living, largely to enable rich and powerful people to kill wildlife for fun. But as I set out at the beginning of this book, I cannot understand the lives of others from within my own bubble. I need to step out and learn. And also I must realise that things change; while they come with baggage, attitudes change with technology and events. Pheasants used to be shot far more extensively, not in the intensively farmed way they are now. But the innovative Victorians popularised breech-loading shotguns, which were a game changer. The faster reloading time meant more birds could be shot, which in turn led to supplementing the number of birds available by breeding them, and by controlling more vigorously the potential predators.

This in turn was curtailed by the arrival of the First World War, which took the keepers to the trenches. Just in time, perhaps, to stop animals like the pine marten being completely eradicated. Now there are far fewer keepers in the United Kingdom, down from 23,000 to 5,000. But the industry has intensified, with a greater demand for birds to be bred to shoot.

My quest to find a willing keeper began, as is becoming the case with this book, with reticence. But I remembered someone who I had seen at many Mammal Society and hedgehog meetings – Jonathan Reynolds. I had always, if I am honest, been a bit intimidated by him; not only did he work for what could be considered as 'the other side' – the Game and Wildlife Conservation Trust (GWCT) where he helped invent the smart trap used to catch mink – he was also the sort of person who would sit quietly, listening, and then drop in a devastatingly astute comment that left some of us feeling rather ignorant. He was, however, always polite and I risked dropping him an email in his retirement.

He was so welcoming. I went to visit his rather isolated home in June 2022, amid farmed land and chirruping swallows, and found that a short coffee extended into most of the day as we found so much common ground. A shared love of photography and music – in fact the only time we really disagreed was when it came down to who recorded the best version of Sibelius's *Fifth Symphony*. While he identified the stand-out version to have come from the baton of Malcolm Sargent, for sentimental reasons I know that the best came from Simon Rattle.

Back to business. I was bothered about the concept of identifying a date to which we were trying to return the ecology of the country and he was very clear, saying, 'It is not a question of turning back the clock, it is about making the best of what we have left. And we have to accept that predator control may be a part of this.'

We then moved on to talk about how 'control' was frequently inadequate. As he said, 'It is not how many you kill that is important, it how many are left.' Which takes us back to the hedgehogs on the Uists; success can never be measured in the numbers of the dead.

'Take the coypu, for example,' he said. 'These huge rodents in East Anglia – and they are big, weighing up to 7kg – were introduced from South America for fur farming. Inevitably they escaped, making their home in the wild, being a pest to farmers, but also enjoying the native vegetation to its detriment. A campaign to eradicate the wild coypu was started in 1962 but failed in large part because trappers were being paid a bounty for killing them. This in effect reduced incentives to remove them all … so it was restarted in 1981 with payment being focused on successful eradication – the focus was shifted from the number killed to the number left – and a bonus was available on completion.'

The successful eradication of the coypu is an important moment in understanding how to approach these complex issues. Though Jonathan also notes that the eradication was also greatly assisted by some very cold winters.

Countries such as France and Belgium, that didn't manage to exterminate coypu before they became too numerous, now suffer continuous damage to levées and other waterside banks. Over time the costs of fixing far exceed the outlay that would have removed the animals.

Jonathan, while enjoying some field sports, is much more of an academic. And while his insights were illuminating, I needed to find someone who was closer to the ground. Someone who was actively keepering. He suggested I reach out to a one-time colleague of his, Dr Mike Swan.

Mike is a senior advisor to the GWCT, and agreed to meet. He did not baulk at my request to find out more of his world, to try and understand what made him tick. And then the stars aligned and I found that a talk I was giving in Wiltshire in October 2022 was not too far from his farm just over the border into Dorset. I suggested the day for a meeting; he invited me for lunch. I responded by saying that he had probably already assumed that I was a vegetarian; he said he would make tomato soup with vegetable stock.

I do not believe it is ever going to be possible to understand people with different views if we do not spend time in their company, preferably physically; though intellectual curiosity can be a good alternative. The aphorism is to walk a mile in someone else's shoes, but I am a bit squeamish about shared footwear. Roman Krznaric has written eloquently about the necessary power of empathy. He has taken his writing out into the real world too, helping to establish the Empathy Museum, where you can literally walk a mile in someone else's donated shoes. Housed in a giant shoebox, the shoes also come with audio stories from the people who donated them, so you can walk a mile in the shoes of a Syrian refugee, a sex worker, a war veteran or a neurosurgeon. Another of their projects is 'A Thousand and One Books', where you get to choose a book from the travelling library, but you choose it from the dedication left by the donor (and then pass the book on when finished). But I think my favourite is the Human Library, where you don't borrow a book, you borrow a person for a conversation – a living book.

One of the best descriptions of empathy come from Harper Lee's *To Kill a Mockingbird* with Atticus telling Scout that truly understanding someone requires that you 'climb into his skin' and go for a walk. And this is what I had to do – however uncomfortable the shoes might appear, or ill-fitting the skin, I had to at least attempt to understand.

The village Mike lives in is a dead end. There was a sneaking feeling of apprehension as I found my mind heading into dark places. What would happen when he found out how different we were? But that is just silly, and as he greeted me with a cheery smile I had already begun to relax.

It turned out we have a mutual friend. Dr Andrew Lack is a senior lecturer at Brookes University in Oxford, where I am a visiting lecturer. I have known Andrew for years; he was a choice of subject when I was writing the book *The Beauty in the Beast* as my robin expert, having just published a gorgeous volume looking at the way these birds have made it into culture and literature. He built that book off the back of the work by his father, David, who really developed the idea of scientific ornithology, including the Lack Principle in 1954: 'The clutch size of each species of bird has been adapted by natural selection to correspond with the largest number of young for which the parents can, on average, provide enough food.'

That might seem an esoteric addition here, but it was one of the few things I remembered from my days as a student.

And it gave Mike and me a good grounding of connection while he ground the coffee beans. He had been a PhD student when Andrew was a postdoc at Swansea, and it seems that the very refined Dr Lack had quite a wild youth ... it definitely puts the violin-playing, pollination specialist in a new light.

I mentioned the demijohn in the kitchen, stuffed with fruit and liquid – was this sloe gin on the way? 'No, this is blackberry whisky,' he said with a satisfied smile. I looked shocked; maybe because I am very fond of whisky and do not like to see it besmirched. 'It's alright,' he added, 'I only ever use cooking whisky. And you must pick your berries early season if you can; they are more acidic, makes for a much better drink. So that is – for a litre of whisky – 1.2 kilograms of blackberries, and 400 grams of sugar.'

He also explained why there was a time after which you should not pick blackberries; you must always harvest before Michaelmas Eve – 29th September – because it was on that date that the devil fell from heaven, landed in a bramble patch, and in fury, cursed it with a spray of urine.

'I would never preach that mine is the one, right, way,' Mike began as we settled into the sunny conservatory with our coffee and away from recipes for liquor. 'I trap, shoot and use cable restraints with the simple goal of allowing the grey partridges to breed successfully. And the pheasants; we don't release birds into the wild here, we create an environment in which they can successfully reproduce.'

This immediately puts Mike into a different league of keeper. The 47 million pheasants and 13 million (non-native) red-legged partridges released each year just for the fun of killing completely distort what is left of Britain's wild.

Mike has had an influence on some of his neighbours too; the Cranbourne Estate has also given up the release of farmed birds. 'Rare as hens' teeth,' he said when I asked if there were more shoots taking on this policy.

Mike does not own the fields over which he hunts; instead he has an arrangement with the owner, which seems a little one-sided to me. Mike manages the land for the benefit of wildlife – including the birds he wants to kill – and in return, gets to go out and kill them.

'This is a manmade world,' he said, 'and one of the consequences of that is we are pushing species to the brink. It is our moral responsibility to stop that happening – like the partridges. Thirty years ago they were all through Dorset, now there are none apart from around here, because of the hard work a small group of us have undertaken to keep them thriving.'

I pull us back to a comment he made, almost in passing, about 'cable restraints'. I ask if that is just a less confrontational term for a snare?

'Oh, these are just a million miles from the snares of old,' he said, without a trace of defensiveness. It turns out he helped design the new snares. 'These have a very low welfare impact on the captured animal, in fact they are used by wildlife ecologists to trap foxes to then attach radio collars to, so they would not be wanting to study a damaged animal.'

He could tell I was sceptical, so he continued. 'Foxes are such an important predator of so much of our wildlife. They are there at unnaturally high levels because of human activity. This means they can impose intolerable predation pressures on all sorts of species, and if we haven't got the means of controlling them we can expect a lot of our declining wildlife to get even rarer and in some cases go extinct.'

This was such an interesting position to find myself in. I really wanted Mike not to feel like it was to be a combative session; I sincerely wanted to know what it is that makes him tick. There was a degree of tongue-biting going on at that moment. The reason there are as many foxes as there are is that they are being sustained by the detritus of the shooting industry. Those 60 million birds do not all end up in the pot; scavengers make many a meal out of them too. And not just the foxes, either; the corvids are also well fed.

'Generalist predators, like the fox', he explained, 'have a wide prey base, so if there are no partridges, there is no problem. But for the partridge, for the relatively short period of time that it is breeding, the presence of the fox is a big problem. Have you heard about the ravens and the tortoises in the Mojave Desert?'

That was a jump, I said. But as was becoming clear with Mike, there was a point to be made.

'By scavenging roadkill, the ravens increased in number. For most of the year there is no problem for the tortoises, but in the autumn, when the young are dispersing, they are easy pickings for the ravens. Now there is zero productivity. There are still old tortoises – shells thick enough to protect them from beaks – and they are still breeding, but no young survive because of the ravens.'

There had been many attempts to dissuade the ravens; Mike described them as 'high-cost, low-return alternatives', but the only thing that helps is killing the ravens.

I took a look into this story, and what a fascinating conundrum. Fifty years ago ravens were rare in the desert. It seems that it is more than just roadkill that is allowing them to flourish – our rubbish is also a great source of sustenance, we have provided water in our gardens, and we have been busy erecting potential nesting spots with mobile phone masts, pylons and billboards. Numbers are up by 700 per cent in the West Mojave Desert in the last 25 years.

Tortoises are slow to reproduce; they are long-lived and well defended so have no need to increase their chances with massive clutches of little ones. They are wise to the desert, where temperatures can reach 60°C, spending much of their lives in burrows, emerging to eat and drink at more Goldilocks times, before the desert night gets too cold.

Ravens are clever; they learn, they wait, and are able to break into the shells of the young tortoises. The resulting pierced carapaces are clear evidence of the predator. Tortoise numbers are down by over 90 per cent since the 1980s and ravens are implicated in up to 91 per cent of the mortality. This is an existential threat to the desert tortoises.

What to do? Because ravens are not without their fanbase. They are magnificent birds, and while I was bouncing around the fields with Mike later in the day, I was thrilling at seeing so many above the land. But for him, they are not a threat – he has his sights set on other corvids.

Back in the desert, efforts are being made to try and control the raven problem. The high-cost, low-returns alternatives that Mike mentioned are things like using drones to squirt eggs with corn oil, so that the adults will continue to sit on them, despite the embryos having suffocated. Oiling eggs is used on many species as a humane way to control numbers.

The U.S. Fish and Wildlife Service have been using this technique to control ravens in defence of the greater sage-grouse. This large, iconic bird of the plains of the western US is already in the midst of politics, as efforts to protect it are held up as an example of interference with the exploitation of the fossil fuels that lurk beneath their range. A range that is already depleted as the grouse is hit by the same sort of onslaught that impacts so much wildlife; clearance for farming, overgrazing and climate change have caused numbers to fall by 65 per cent since the mid 1980s. And also in the mix are, unfortunately, ravens, whose numbers have risen threefold in recent years.

Ravens have followed people, have been inadvertently supported by people and are also not aware of the fondness those people have for the grouse, so, when the young pop up as easy prey, the ravens are ready. Whether it is tortoises, sage-grouse, or least terns – another species that has come under the gimlet eye of the raven in southern California – conservationists see that something has to be done.

Obviously the story is complicated, because it would not be ecology if it were not complicated. The ravens are a symptom of many other threats that these species face, but are the symptom that is most accessible, the one that we can do something about. Removing the threat is the obvious answer. Which is clearly code for killing.

Killing ravens presents problems. They are intelligent. Shooting is tricky, nest removal tough work, and leaving poisoned eggs around risks hitting non-target species. And evidence from other grouse species, such as willow ptarmigan and black grouse, shows that simply killing the corvid does not result in an increased number of grouse.

There is the additional problem that there is no end point. This was something that Mike brought up a great deal – a successful control of a wildlife-on-wildlife issue is one that ends. If it simply runs on forever, it cannot be considered a success.

Alternatives to killing ravens have been explored, and some of them are really boring. It is about modifying human behaviour so as to not support the ravens to such an extent. Rubbish attracts them; where there is litter there is probably food. Where there are roads there is roadkill, which needs to be cleared up quickly. Industrial meat production generates waste, especially around where cattle give birth. Leaving that for the scavengers helps ravens. As does the creation of nesting points on billboards and pylons.

Some interesting work has been done looking at exploiting the intelligence of the birds. The chemical methiocarb causes ravens to feel sick, and as such has been used in an attempt to develop aversion to least tern eggs. But Mike was sceptical about overcomplicating what should be a simple solution.

After a bowl of really delicious soup, Mike offered to take me for a tour around 'his' estate. He is clearly proud of the work he has done to get it to the state where he can invite a few friends to join him on what he describes as 'an armed nature ramble.'

'I am a pot hunter – I shoot with eating in mind. For 40 years I have been working with the GWCT, advising shoots – driven shoots, where birds are bred for the purpose and then beaters chase them into the air and towards the guns – but I very rarely go on such things. Actually my favourite recreation is coastal wild-fowling. I love to be out in the wild, just me, or maybe a friend as well. Or woodcock in the depths of winter, or bog-hunting snipe. I love to be in these places, though I have found that I am getting less hungry for the kill as I get older. I shoot one or two and then stop and have a cup of tea and just enjoy the space around me.'

He paused before continuing. 'You know, the biggest buzz I get is when I pull off a shot and the bird is dead in the air.'

I wondered whether Mike got a similar thrill from killing the animals he sees as competition. His answer was intriguing.

'Oh, I have no particular joy in killing a crow or a fox; maybe a sense of satisfaction, but no thrills from these kills. My motivation is to produce enough wild pheasants for me and my friends to shoot the makings of a dinner party. Maybe eight of us will take 20 pheasants, a woodcock or two and maybe a duck, for feast and freezer. The game is social glue. And one thing it is not is a sport to test us – if that is what you think you are doing then you are doing it wrong.'

This required some explanation, and the point that Mike was making seems very reasonable. He described how there is a long distance between certain death and no harm when you fire a shotgun, for example. And there is always the thought that maybe you will be able to do it from a little further away or in poorer light than last time, but this is not a target you are aiming at. If it is not killed outright, you have failed.

It was the unapologetic nature of Mike that was growing on me. At no point was he hiding what he did. And yes, going on an armed nature ramble is nothing that I would want to do, but it makes sense to describe it like that.

Some years back I was doing a travel piece for the *Telegraph* and ended up in Tobago – I know, a tough job. But I was not covering beaches, I was off walking into the middle of the island, looking for waterfalls with a local guide. Soon being picked up and getting off the more main roads, I pointed to a bird in a tree and asked him what it was. He pulled up and got out, went to the boot to get binoculars, I thought, but no, he emerged with a gun, proffering it to me. We had very different expectations for that nature ramble.

Mike is one of the few people I have really talked with who enjoys shooting. But who is the odd one out here? I know he is not alone, yet I so rarely come across anyone who thinks like him … and yes, we have all found our lives encapsulated within the social-media-facilitated echo chamber.

'There are over a million shotgun licences alone,' he explained. 'Now, many of those will be owned by farmers who may not use their guns very often, but the British Association for Shooting and Conservation has a membership of around 150,000. This is a much more normal thing than you might think.'

We have drifted off topic a bit. Yes, there is a lot of killing done for pleasure, but I am most interested in the killing that is done to allow his version of fun to take place.

'Okay, I think the most important idea that you need to understand is that I am *not* interested in predator control.' Mike is clearly well versed at stringing along people like me. He paused just long enough before clarifying. 'My mission is predation control. If my partridges and pheasants are doing just fine then there is no need for me to kill corvids and foxes. So, my Larsen traps – I am very specific in how I use them. There is no point trying to kill all the crows and magpies that visit this land. That is the Victorian style of gamekeeping, and there are still folk out there who get a kick out of joining the 'ton-up' club, when they have killed 100 magpies in a season.'

Mike's approach is based around ecology and behaviour. The crows, for example, will come in and create a territory and learn where the nests of pheasants and partridge are and wait. If you take out just these territorial crows, and any that try to replace them, you are left with just the non-territorial birds that are visiting the area on the off-chance and who have much less impact on the game bird population. These do not predate the nests in the same way. Additionally, and obviously, if the problem is nest predation, when the nesting season is over, you stop trying to kill the crows, as there is no point. This way Mike averages about 20 crows per year.

The corvids get such a bad press. This had left me confused as I adore them – their intelligence, their capacity to cope with humanity. And then I saw a magpie killing a fledgling blackbird. It was in the park behind my house and I was out walking the dog when I heard a commotion and saw the magpie as it pecked, fairly ineffectually, at this poor youngster. And it dawned on me: they are just not very good at killing. There is little upset when a peregrine stoops on a pigeon (unless it is one of your racing birds), because the peregrine is a masterful predator and in a second the bird is dead. The magpie was clumsy and that gave the impression of cruelty.

I expressed this idea to Mike, but he took it further. 'Our problem with the corvid is deep-seated; it goes back through history. I reckon the bad relations started with warfare – human bodies being scavenged by the birds that were commonplace. That has to change your attitude towards them. Add to that their penchant for stealing bright and shiny things. I believe that for many people it is in the same sort of deep place as our reaction to snakes – I know that I am very unlikely to be injured in any way by a snake, but I still start when I see one.'

Earlier on Mike had expressed his desire to be as humane as possible. I had to ask him more about his use of snares ... or as he prefers, cable restraints and the Larsen traps. Both of these have been regularly demonised by opponents of this lifestyle.

'The reason I use these methods is absolutely due to my passionate desire to cause minimal suffering.' He paused as I was obviously struggling to keep a straight face. 'Look, when I shoot at a fox, there is a risk that I will injure it; that is completely unacceptable.' And for the first time I see him getting really emotional. This is important to him, that he is able to restrain the fox or crow and then kill them cleanly.

'Look, the weather is good, so let's go out and let me show you why I do what I do.'

A short ride in Mike's car took us to a barn with the most delightfully beaten up Land Rover I have ever seen. There was bailer twine holding various pieces together and the passenger window was held up by a wooden wedge. Conversation was obviously a bit harder now, but we persevered.

The first thing that was clear is that this was no normal farmland – there were strips along the hedgeline, and across some of the fields, that were what more conventional farmers might describe as rank. We were bouncing along one of those now. 'These are the beetle banks,' Mike shouted over the engine noise. 'Look!' He pointed as the first of many hares darted off from under our wheels, but it was not the hare that had caught his eye, it was a bird – a yellow wagtail. 'That is the first I have seen here in 20 years,' he said with a smile of deep satisfaction. These birds need insects to be thriving so that they can live. The hares kept coming thick and fast. I have never seen so many on one outing.

And then up floated the ravens. To begin with I had the perspective all wrong and, in a moment straight out of *Father Ted*, assumed they were crows or rooks but further away. But no, these majestic birds seemed very much at home in this really quite unwild farmland. Mike was not bothered; there were now five in my view and I was getting excited. He was not worried about them in the slightest. He is a fan, in fact, as he reckons they help control magpie numbers. And he has seen 12 in one flock – what an unkindness!

Moving through some unharvested barley we spotted two roe deer, a female pheasant, yellowhammers, goldfinches and a sprinkling of one of my favourite birds: skylarks. This was a high-speed safari, well, not exactly high-speed, but we were seeing a lot in a short space of time. I was excited by these puffs of nature that sprang into sight, and I was grinning for much of the next hour or so of the journey, despite a growing sense of melancholy. Even what I was seeing here was just a fraction of what there once was, and that hurts, every time.

Aldo Leopold captured this sense back in 1949, in *A Sand County Almanac* he wrote,

> *One of the penalties of an ecological education is that one lives alone in a world of wounds. Much of the damage inflicted on land is quite invisible to laymen. An ecologist must either harden his shell and make believe that the consequences of science are none of his business, or he must be the doctor who sees the marks of death in a community that believes itself well and does not want to be told otherwise.*

The reason so many people miss this pain, until it is too late is due to shifting baselines – the idea that we tend to think of a healthy ecosystem as something with which we grew up, failing to recognise this as an already enormously degraded state. So each generation strives for a diminished goal, our baseline for what is 'right' slipping until we are lost.

More roe deer kept appearing, though we could have been chasing some of the same ones around the fields. I asked Mike whether he ever went out stalking deer on his armed nature rambles. Which was stupid, as you don't ramble after deer, you sit and wait, I learned. 'We have two deer here, the roe you have seen and the fallow we will see soon. I hate the fallow.'

That seemed a bit strong – the fallow deer that inhabit my regular stomping ground in the deer park at Magdalen College in Oxford are gorgeous. But I am thinking of the aesthetic; Mike is thinking in very different colours. 'Roe are much easier to kill, and much nicer to eat,' he said. 'The roe tend to be in small groups, and are themselves smaller, so easier to deal with once I have shot them. But the fallow – as they are in a herd, you shoot one and there are then 19 witnesses who are much more nervous. But I have to keep the fallow numbers down, let me show you.'

We bounced on some more. I noticed a very yellow leaf get caught up in the breeze, but it was a little early for autumn, and there was no breeze and Mike stopped, and started for his binoculars. 'Clouded yellow butterfly,' he said with a satisfied smile.

Coming to the edge of the fields Mike slowed and pointed to the hedgerows, which had a clear browse line. Nothing grew beneath about 1.4m – stripped by the fallow. This massively decreases the biodiversity and bioabundance potential of the habitat. But there was more. Looking into the woodland it was now easy to see, knowing what I was looking for – the impact of the deer. Ground cover was minimal, and no foliage was growing on the trees under that 1.4m height.

The problem is obvious: the deer are living without fear. The predators that used to keep populations in check have long since been killed off by us, leaving humans as the agents of control. Without a control on the number of deer, there will be a continued decrease in the ecological value of our land.

'You know how everyone cites the Knepp Estate as this great example of rewilding,' he said. 'Well, you should ask them what they did about their deer.' I have seen photographs of an archway built at Knepp from fallow antlers - though these might have been discarded rather than shot. Now the deer are shot and the meat sold.

The impact of herbivores is often overlooked when wildlife management focuses on predators and scavengers. Knepp is far from the only place that has had to take action on deer to conserve the landscape. The absence of wolves is sorely felt. So humans have to step in and, while not howling into the full moon, at least take over the predatory role. Across estates in Scotland, such as Glen Affric and Glenfeshie, the results of culling have been dramatic. Fewer deer allow the vegetation to grow – Scots pine and juniper regenerate. Those trees that remained used to just be the old – they were arboreal retirement homes – now youngsters are poking up into the light. To achieve this, deer numbers have been reduced from 50 per km^2, down to just one or two.

'People have to get over the emotional hang-up about killing,' he said as we peered into the patch of trees, noticing for the first time the fallow deer amongst the bare trunks, keeping a very watchful eye on us. 'It gets me frustrated, people who happily eat processed meat, killed by someone else, but get upset when I try and improve the land by killing a deer, and eating the meat. I think we could all benefit from less meat being eaten. Now there is no way I will be turning into – what did you call yourself? A vague-an? But if less were eaten, well, the welfare of what goes on in battery farms and slaughterhouses is a world away from a clean shot of a partridge or a roe. And that is not even looking at the climate-change impact – the oil for fertilisers to grow cows.'

This is proving to be a strange point of agreement. When I was doing some investigative work with Compassion in World Farming what fascinated me was how the organisation at the time was run mostly by vegans, yet they were not pushing that agenda; they were looking for people to eat fewer animal products and to choose products that were of higher welfare standards. The phrase 'meat as a treat' was one I tried to push on them to take. Just imagine if millions of people simply ate less meat – that would have far more impact than a few thousand eating none at all.

The lack of noise from the Land Rover was refreshing and we both got out to look more closely at the browse line and the lack of ground cover. I was forced to agree with Mike – the deer clearly have an impact. Taking advantage of the peace, I thought it time to broach some of the more contentious issues.

The disappearance of birds of prey, especially species such as the hen harrier, seems to correlate very well with shooting estates. Notwithstanding 2022 being held up as a great year for breeding success in harriers, they are still being killed. It is very hard to get direct evidence, as vast swathes of the countryside are off-limits to members of the public due to the rather silly trespass laws of England and Wales. At least Scotland has opened things up (without the collapse of civilisation, as prophesied by the more introverted billionaires). But there are moments when brave investigators, either through infiltration or careful positioning of cameras, are able to catch these crimes. How does this make Mike feel?

'This always leads to a very vociferous debate,' he said. 'But the noise comes from the fringes – from the extremes of both sides. And this is a problem as there are many people in the middle who are quiet. I write for many publications and it is so hard to try and explain something this complicated in a soundbite.'

But surely there can't be much nuance when birds are found to have been poisoned, I ask.

'Pesticides can kill birds of prey, of course. But there are three different ways this can happen. First, and very rare, is if they are used properly. Then there are cases where they are used improperly and you get inadvertent fatalities. And finally you get deliberate cases of abuse, which I argue is very rare as well. Most cases will be from mistakes.

'Look at the story of those white-tailed eagles in Dorset,' he continued. 'An enormous story – these birds have been reintroduced to the wild and were found poisoned. But I am sure it was accidental. It is just that it blew up, and then the MP Chris Loder weighed in and just upset everyone. What a fucking idiot.'

Chris Loder was quite the accelerant for this bonfire of common sense. He declared that Dorset was not the sort of place for these birds to be and went on to heavily imply that the police should not be prioritising investigating how they turned up dead on a shooting estate with seven times the lethal dose of the rodenticide brodifacoum in their bodies back in January 2022. It was also revealed that Loder was the recipient of a considerable donation from a shooting estate in his constituency, though whether it is the one at the centre of the eagle storm is unknown.

The fact that so much is known about the fate of these two birds is down to the amazing work of Dorset Police's wildlife crime team, headed by Claire Dinsdale. But while she had collected evidence that warranted a search to be carried out on the estate on which they were found, Dorset Police issued a statement in March stating that the toxicology results were inconclusive, 'and it has therefore not been possible to confirm that any criminal offence has been committed ... As a result, no further police action will be taken in relation to this report.'

To add insult to this injury, the Rural Wildlife & Heritage Crime Team was disbanded and reformed as the Dorset Police Rural Crime Team – while Claire Dinsdale was on long-term sick leave. She eventually moved to the National Wildlife Crime Unit.

Campaigners have not left this story alone; a series of Freedom of Information requests were submitted, as were questions over the involvement of the local MP in the decision to drop the case. Dr Ruth Tingay, who writes a blog called 'Raptor Persecution UK', concluded that 'Dorset Police's refusal to answer FoI questions simply left these concerns hanging in the air like a bad smell.'

Dr Tingay also collected the tweets that Claire Dinsdale had put out (before they disappeared) in which she said, 'The eagle case was shut down prematurely in my view & the planned multi agency search I had arranged was cancelled by a new boss with no understanding of wildlife crime and a very senior officer within days of an MP's rebuke & threats on police funding…'

This is just one case in many that does leave the casual observer with a sense that not all may be well with the way wildlife crime is policed. And it was obviously a case about which Mike had strong feelings. But he had an interesting idea as to why this debate has become so polarised.

'It's all down to the way Tony Blair's government introduced the ban on fox hunting,' he said. 'You see, what this did was create a great sense of betrayal in the countryside – it was a turning point in country people's respect for the law. They saw it was an unjust law, one that could not be enforced, so stepped over the line.'

Mike's thesis is that this marked a moment from when the usually law-abiding and conservative souls of the land felt unbound by the law. However much the rural lobby may feel slighted by the democratic wishes of the majority, that is hardly a good excuse to go around killing protected wildlife. 'The honest operators should not have to suffer because Natural England does not have a system in place to catch bad apples,' Mike justified.

I really liked Mike, but felt that we had hit upon an area of fundamental disagreement. I feel that honest operators would be well placed to indulge in a little self-policing.

It was time to head back so we clambered into the noisy machine and bounced back to where he had left his car. As we pulled up I noticed a bunch of wire cages in the back of the barn. I asked if these were the Larsen traps he had mentioned. He reached down a couple of cages to show me, with the help of a rather bedraggled cuddly toy crow, how they work.

'They were invented in the 1950s in Denmark, but I have modified them a bit,' he said, quite proudly. Their design uses the gregarious nature of inquisitive corvids against them. In one cage sits a captive, a decoy bird, which is where the toy came in for this demonstration. This is considered an intruder by resident crows or magpies, who will attempt to drive it away. When they get close and enter the trap, the door shuts.

But why not design a trap that just kills the target, saving them the stress of spending hours in a cage?

'Well, for one, it is not that stressful, the birds should have shelter and food. But the main reason is that with a kill trap you always risk killing the wrong species. Though to be honest it is very rare that a non-target bird gets caught.'

The Game and Wildlife Conservation Trust, in its guide to the use of these traps, states, 'It is important to remember that in today's countryside, the future of shooting depends on game management being conducted responsibly and professionally and in a way which delivers wider environmental benefits.'

Which would be great if there was enough trust between members of the public and the keepers. But repeated cases of abuse make trust-building hard. And the cases which do see the light of day are widely accepted as being 'the tip of the iceberg', according to Dr Ruth Tingay, who told me that: 'This is due to the difficulties of quantifying wildlife crime, particularly where it occurs in remote areas where the circumstances severely limit the number of potential witnesses. Indeed, what is usually found is the aftermath of a crime, as opposed to the witnessing of a crime in progress. Added to this are social and cultural pressures preventing whistleblowing and inhibiting certain sectors from reporting wildlife crime incidents. For some wildlife crime, notably raptor persecution, its extent and scale can be determined by other sources of evidence. Long-term scientific data have emphatically shown that raptor persecution is so prevalent, particularly on land managed for driven grouse shooting, that it is having population-level effects on some species, e.g. golden eagle, hen harrier, peregrine, goshawk and red kite. The Scottish government has finally accepted this evidence and is in the process of regulating and reforming grouse moor management with new legislation to tackle these crimes but the Westminster government remains wilfully blind, largely due to vested interests and the hugely influential landowners' lobby.'

The onus, I believe, has to lie with those who want to continue killing for pleasure to demonstrate that they are in every way so utterly scrupulous. It should not have to be the responsibility of the authorities and members of the public to police the activity.

'The trouble is', Mike concluded, as we returned to his home and I was getting into my car, 'that we have too many people with too many emotional hangups – those hangups need to be left out of the equation.'

I enjoyed our conversation enormously and hope that Mike would be willing to have more. From an entirely utilitarian point of view, his position makes perfect sense. If we are to measure success as having more grey partridges, and wild, breeding pheasants, then the use of the Larsen trap generates a healthy result.

This is the big disconnect for me when it comes to trying to understand Mike. My inability to set aside my emotional reaction to killing for pleasure interferes with my ability to see the spreadsheet - the balance of good and bad - as he does. Which leads me to believe that a spreadsheet that ignores emotion is going to be incomplete for many people.

Mike wants to keep the disturbing emotions away from his sport. Yet as we travelled around the estate, it was clear that what drives him is not just a utilitarian calculation – there is love and passion for what he does. He loves nature; he was clearly filled with the same joy I was.

The difference between us is so stark. It reminds me of days at school when I was perceiving that there was no god and that we are all alone. I spent hours in conversation with believers; in fact this has been rekindled by my son's life as a chorister at Magdalen College in Oxford. I spent five years going to evensong – up to five times a week. Fortunately I love the music and having my little boy there, singing like an angel, boosted the experience to the far reaches of my heart.

I got to listen to an awful lot of the Bible. And the prayers and sermons too. To the embarrassment of my wife I would, however, listen and take notes. Then, as we filed out of the ancient candle-lit chapel, I would make a beeline for the dean of divinity. For much of my time there it was Jonathan Arnold in the hot seat – fortunately he was blessed with good humour and patience! Often he would brush my questions to the side with a 'well, it is awfully difficult to generalise'.

My point is, Jonathan is a wonderful and intelligent person, I enjoy his company. We are both socially liberal, keen on a better world for us all, desperate for the war on nature to cease. But we have built our shared desire for a kinder world on completely different foundations.

And it is the same as Mike: we both love nature. Do we need to be reconciled to work together? The trouble is that Mike is a rare example from the gamekeeping world. I am sure there are others who have such high standards and a strong ethical base, but they are not talking to people like me – nor offering such fine soup.

Cocaine hippos

No, this is not another way of describing drug mules, and this is not a competition to tell the strangest stories. If it had been set in Florida, it would already have featured in a book by Carl Hiaasen.

In the 1980s, drug dealer and all-round bad sort Pablo Escobar illegally imported exotic animals, including four hippopotamuses, into Colombia for his private zoo. After he was shot dead by police in 1993, the hippos were allowed to roam free and made themselves at home in the Magdalena River basin. Here, their numbers began to increase to around 150 – that is 150 3-tonne specimens who were not meant to be here.

These animals are notorious as some of the most dangerous in their African homeland, killing around 500 people a year. So far there have been no fatalities in Colombia, but ecologists are very concerned as to what these aliens are doing to the ecosystem. The hippos' new home is also home to West Indian manatees, Neotropical otters, spectacled caimans, and turtles, who are being displaced as vegetation is consumed by the hippos. Additionally the change in the nutrient content of the water has led to toxic algal blooms, killing aquatic fauna.

So in 2009 a plan was launched to cull the hippos. But as photographs emerged, a national outcry brought that to a halt. In addition there was a competing view expressed by conservationists who questioned whether the presence of this megafauna might not be a boost to the ecosystem – though this was based on a best guess from how they integrate into their more traditional African home. Possibly grasping at straws, some suggested that the animals could fill the place of extinct giants such as *Toxodon*, offering a form of Pleistocene rewilding.

The next plan was to at least stop the numbers increasing. The first effort, castrating a male in 2017, cost around

$50,000 and was not simple. So now there is a plan to sterilise animals using an immunocontraceptive vaccine. And the latest move: ship 70 of them to sanctuaries overseas, with 10 scheduled for Mexico and 60 going to India. At a cost of $3.5 million. And if that is not the beginning of another disastrous ecological saga – or better still a big-budget movie – I don't know what is.

CHAPTER 9

New Zealand

Most of the people I have talked to about killing for conservation have mentioned New Zealand – and for very good cause. Over the last few centuries Britain, along with many other European nations, decided that what was once someone else's would now be theirs. But the age of Empire – the time of colonial rampage that set in motion the revolutions of industry and society that have led us to the point where the climate and ecosystem are collapsing – had another impact as well.

In his fascinating book on the Anthropocene, *The Human Planet*, Simon Lewis argues that this human-induced geological epoch could originate from the Columbian Exchange; starting at the end of the fifteenth century with the 'discovery' of the Americas. The more predictable markers of the nuclear age – the industrial or agricultural revolutions – have had, arguably, less impact on the ecosphere than the moment at which we started to ship species from continent to continent. We inadvertently recreated Pangea.

While Lewis was referring in large part to the shift of agriculturally important species, these were accompanied by many other hitchhikers that have gone on to have a staggering impact on the ecosystems into which they arrived. We now know the innocuous earthworm has transformed parts of North America, for example. But here I want to look at some rather more dramatic introductions, accidental and deliberate.

In particular, in New Zealand. We have already seen how complicated things can be, and the trouble that 'my' hedgehogs can cause. But they are only a small part of a much larger problem.

In 2016 New Zealand's then prime minister, John Key, announced, with great fanfare, a new plan: 'Predator Free

2050'. Rats, stoats and possums kill 25 million native birds annually, costing the economy $3.3 billion every year, he declared. This ambitious project would see every single part of New Zealand free of these species by 2050.

Given the amazing amount of damage our careless releases have caused, this seems like a worthy goal. And while hedgehogs are not caught up in this particular part of the work, as we have seen they are most definitely in the cross hairs of the conservationists. And while some of us may baulk at the idea of so many animals being killed, alternatives are hard to imagine (though I wish with all my heart that we could find some).

To get an idea of how New Zealand got to this point, it is worth looking at some of the history, both social and natural. Published in 1984, the book *Immigrant Killers* by Carolyn King is deceptive; it looks intimidatingly academic but is written with the lucidity of a fine nove.

Much of what makes these islands so special, King suggests, is down to the vagaries of tectonic plates; New Zealand slipped its moorings from Gondwana around 80 million years ago and headed out on its own. And there, undisturbed, life flourished until the grand cataclysm (no, we have not got onto us humans yet) of the Pleistocene ice sheets, which wiped forest from most of the land and caused mass extinctions of animals.

The great untouched beauty of New Zealand is actually quite young, emerging as the ice retreated around 10,000 years ago. Only seeds on the wind and sea, floating alongside birds and insects, could help boost the relics of life that had hidden in the ice-free north of North Island.

But amazing things did flourish. With the island being mammal-free until around 800 years ago, birds and insects evolved into many typically mammalian niches. The larger wetas, amazing insects whose genus name, *Deinacrida* means 'terrible grasshopper,' took on the role of mice; large flightless birds, the moa, grazed with beaks and not teeth; and the predatory Haast's eagle, hunted the moa by sight – meaning that potential prey learned to stay still, hiding until danger had passed. Which is great until you bring a host of scent-sensitive hunters to the land.

An additional feature of isolated evolution with limited predation is a tendency towards animals getting larger; birds becoming flightless and a shift towards slow K-selected breeding strategies. This is important – it means that as there was a good chance of offspring surviving, more effort was put into rearing fewer of them. The alternative, r-selection, focuses on producing a greater quantity of young, with limited parental engagement. Slower-breeding animals do better in calmer environments; if there is regular tendency to drama, breeding fast stands you in better stead.

Around 800 years ago the Polynesians arrived in New Zealand with dogs, chicken, pigs and rats, starting the new wave of destruction, both from the hunting humans and their sidekicks. All nine species of moa were exterminated. This happened very quickly with a very low density of people on the islands. It is believed that the moa were functionally extinct within 150 years of the first arrivals, by which time there were only around 2,000 people – that is all it took. Remember – *Homo occisor* is our origin story.

The reason I want to put this information here so baldly is to try and resist the temptation to romanticise Indigenous cultures. We, as a species, do bad things when we arrive in new and beautiful places. I think that is a reasonable value judgement to make.

While the Polynesians, as they evolved into the Māori, started the extinctions, they were rank amateurs compared to what was about to come. Europeans arrived in the region first in 1642, but were rather rebuffed, never making it to land. That had to wait for Captain James Cook in 1769. The first notable inroads into the native wildlife came from the sealers, alerted to this new bounty to be found on the south of South Island. Starting in 1792, 4,500 skins were collected after the first 10 months.

This was not the first time seals had been hunted; the moa hunters had also killed seals, but that had been for subsistence. The involvement of the market in wildlife exploitation will always tend towards disaster, as the rate of renewal of the target will rarely match the rate of return to be obtained from liquidating the target. Carolyn King described it eloquently: 'The sealers worked with "reckless efficiency" to put themselves out of business.' And by the 1830s it was no longer profitable, unless supplemented by whaling, farming or trade.

It did not take long for the whaling industry to become undone by further 'reckless efficiency', but the destruction did not stop there. There was a tree, the northland kauri, which generated strong and long straight timbers for ships. This was extracted, with no thought to the wider cost to the forests, leaving as King said, a 'melancholy scene of waste and destruction.' There were estimated to be 12,000km^2 of kauri forest before this onslaught began. About 4 per cent of that remains today.

Not to be outdone, the mineral adventurers got into the mix, with gold rushes in 1861 and 1865 adding to devastation of the land – and this again was another short-term boom for some.

Setting this scene is, I think, useful to give some idea of how dramatic the changes have already been to these islands in such a short time.

Now we add into the mix the unintended consequences of the addition of an array of new species, and you can see why this problem is so serious. The laying waste that happened has happened, and has been cemented with industrial agriculture, making New Zealand – despite what the tourism adverts and the *Lord of the Rings* films suggest – a very nature-depleted country.

This has made what was left even more vulnerable to the arrival of predators. There has also been a massive impact from plants. But I am not going to go into those in any depth, as this would then result in a book of unmanageable proportions.

Carolyn King has dug deep into the archives to chart the progress of alien species. Our first deliberate contributions to New Zealand in the late eighteenth century were sheep, goats and pigs; rats and cats were stowaways. Then followed the Australian brush-tailed possum in 1837. The fur of this pretty marsupial was considered valuable. Domestic rabbits arrived the next year, then followed red, fallow and sambar deer; hares and wild rabbits; and, obviously, hedgehogs.

The rats and dogs that came with the Polynesians were some of the first animals to be wiped out by the European invaders – the Pacific rat, the kiore, was rapidly outcompeted by the brown rat. Kiore was a deliberately introduced species as it was considered food, not pest. The Māori speculated about whether they themselves would be exterminated by the Europeans, in the same way their rat had been eradicated by the European rat.

Rabbits were imported by Europeans as a source of food and fur, but by the 1870s they were having a major impact on the ability of the native flora to flourish on both the north and south islands. This meant that the native fauna who relied upon the plants were also impacted. The reaction was to find a way of controlling these rabbits. In retrospect, the introduction of stoats and weasels in an early attempt at biological control seems a little bit thoughtless. Stoats are now one of the targets of the Predator Free programme, having caused localised extinctions of iconic birds such as the kiwis and the kakapo.

It is easy to become fixated on the conservation impact of predators – after all, the killing of native species is the most dramatic result of new arrivals. But we should not overlook the impact of the herbivores. The game species introduced to New Zealand, deer in particular, found an environment untroubled by similar browsing and grazing species, the moa having been wiped out already. The complication is that these animals have utility for many, in that they are part of the recreational hunting industry.

In fact, the language used by the authorities about deer is markedly different to that used for predators – who are characterised as pests to be purged, or that the effort is a war against deadly aliens. The Department of Conservation says on their website: 'Although they are not a native species, deer are valued as a recreational, cultural, and economic resource. They are hunted for sport and to harvest venison.'

Even when it comes to a recognition that all is not perfect, the words used betray a very different ethic. 'But, unless the numbers are managed, wild deer can damage the environment by selective foraging. Large numbers also force deer to compete more for food, decreasing their size and quality. This affects hunting and venison harvest.'

There are now seven species of deer established across the islands: red, wapiti, sika, sambar, rusa, fallow and white-tailed. The repeated introductions show how slow we are on the uptake when it comes to the potential for these animals to cause disruption.

In the rather tougher-to-reach parts of the country there is helicopter shooting – there are some fascinating videos to be found on the internet if you want to witness high-adrenaline wildlife control, or 'venison recovery' as they call it. The videos show a few misses, but lots of clean kills. I do worry about the shots that don't kill – which I very much doubt would ever make it out of the edit suite.

To get a deeper insight into what is going on I returned to my 2022 conversation, which had started on hedgehogs, with Chris Jones from Manaaki Whenua. I had suggested that the difficulty with his work was that ecology was so complicated. He was quick to correct me: 'Wildlife management is complicated, ecology isn't, that is just science.'

And this is an important point, one that I think I have a tendency to forget. Studying the way that the web of life works is definitely hard, but it really only teeters on the edge of the impossible when you introduce people into the equation.

Which are the hardest problems for him to solve, I wondered.

'Cats,' he said, with no hesitation. 'Free-ranging cats on islands, not just New Zealand, have caused, or contributed to, 14 per cent of the modern bird, mammal and reptile extinctions recorded by the International Union for Conservation of Nature (IUCN) Red List. The problem is not that they are particularly difficult to control, but because of the way they are woven into our lives. The species exists on a continuum. From the totally feral cat that most people would be happy to see killed, humanely and economically, of course, all the way to a domestic cat who is a pensioner's dear friend. They are the same animal and can cause the same trouble but must be treated in different ways.'

He went on to introduce an idea that I really think is important and often overlooked. We all have different value sets, and this impacts our judgement when it comes to the complications of wildlife management. Some people have a natural love of wildlife that is outside logic or reason, some people have a complete lack of interest, which I would argue is even more outside logic or reason. But these are my values coming into play – in fact this is a slightly less pejorative way of describing what I called my prejudices.

'Of course', Chris continued, 'values reach beyond just the personal. There are also geographical values too, so the same species, whether feral or domestic, has a completely different impact on our native wildlife if it is in precious conservation land, farmland, the peri-urban space or in the middle of a city.'

What this is creating is a need for a matrix of concerns to be addressed – it adds dimensions to an already complex problem.

'Oh, this is just the start of it,' Chris continued with a wry smile. 'Then there are the economics – not just whether the state can afford to undertake control, but also the domestic economic situation of the individuals with whom we are trying to work. Because while it might seem easy to argue that cats should all be neutered, not everyone can afford that. And if neutering is not done – well, if it slips below 80 per cent of the strays in the area, then you simply end up doing it forever.'

This is a theme that has kept recurring. Wildlife management clearly needs to be done if we as a society have decided that certain species should be saved from extinction. But it can't go on forever – there has to be an end point. Which is why the start of the Uist hedgehog work was so flawed; it was taking so long that the overall number of hedgehogs was possibly not decreasing at all. Which means all the killing, and any suffering associated with that killing, was pointless.

'This is why the Predator Free 2050 programme, that I am involved with, is so important,' Chis said. 'It is enormously ambitious because we are attempting to not be left with ongoing management. That is what would happen if we did not aim for the stars – in fact this was described, at its launch, as New Zealand's moonshot.'

The aim – the moonshot – requires more than just money being spent on doing the same things again, but at a bigger scale. As Chris told me, there is research to be done.

'We are developing novel tools and the dream is to come up with species-specific toxins. Now, because of where we are, we have been able to liberally use the poison 1080 as it really only affects mammals, and there are no native terrestrial mammals. But ideally we would like to do better: to find something that works on mustelids, something else on rats and then something for the possums.'

The most exciting work Chris described to me was the dive into the genomes of the species – 'genome mining' – looking for a part of the genetic code of each of these animals that codes for a metabolic pathway that is essential to life … and then design a drug that targets that pathway.

'It is just like how we are tackling disease', he explained, 'whether it is creating vaccines or cancer cures, the genomes of these species hold so much potential.'

This is clearly where Chris's heart is at – to try and find magic bullets. Leaving aside the potential for complicated and unexpected consequences from such novel experiments, I was interested to know why he was not just amplifying what they already use and know works – trapping?

'Trapping in a park, or small and constrained area, is feasible. But, if we were to expand that effort, we would be exceeding the GDP of our country many times over. It is just not economically possible. And we have found that in all target species, you do get individuals who are trap shy, and as the aim is eradication – a once-and-for-all attempt to return New Zealand back to something more like it was before the Europeans arrived – we can't leave any behind.'

I had read that scent lures have been improving, and that maybe they could be used to at least create barriers to the invaders? But Chris pointed out that most of the scents out there seem to be to some extent attractants – he was surprised to find that the smaller mustelids were drawn to ferret scent for example. 'You might be surprised too,' he continued. 'The best attractant to bring hedgehogs to a trap is ferret, closely followed by dried salmon. They really like something stinky.'

Predator Free 2050 had been going for six years when we talked and the moonshot is being considered in the light of reality. 'We are moving towards a more objective and strategic approach,' Chris explained. 'It is impossible to do the country all in one go, so we have been doing landscape scale modelling, working from areas that are already clear and expanding those, or work methodically east to west for example. We are operating an adaptive model.'

When we had first met, Chris had talked of 'islands'. New Zealand has been fortunate in that a number of its most vulnerable species have retreated to, or been evacuated to, predator-free islands. But to start with I was interested in the mainland islands.

These islands feature one of the problems I encounter with my work on hedgehogs – habitat fragmentation is a curse, creating isolated pockets of land that are too small to support a viable population. Obviously this is a problem that particularly affects terrestrial animals. But in New Zealand these artificial 'islands' present an opportunity.

'When you come and visit you must go to Zealandia,' Chris said. I should add this was not his first bit of work on behalf of the NZ tourism board – he clearly sees a weakness in me. 'Zealandia is the world's first fully-fenced ecosanctuary,' he said. 'Some ecological projects have a small time frame, but not this one. It has a 500-year vision to restore this Wellington valley ecosystem to as close to the prehuman state as possible. The 8.6km fence encloses 225 hectares of forest and water. The project has already reintroduced 18 species, many of which had been absent from the mainland for the last 100 years. You can now see tūī, kākā and kererū, birds that were recently so rare, in and around central Wellington.'

I am unfamiliar with New Zealand birds, so needed some guidance as to what was now in place. 'The tūī is a large honeyeater – a boisterous blue, green and bronze bird, a bit bigger than your mistle thrush. The kākā is a medium-sized parrot, they nest in cavities and sit on eggs for up to 90 days, making them really vulnerable to attack – stoats and possums killing adults, chicks and destroying eggs. The kererū is the New Zealand pigeon. This animal is considered a taonga – a cultural treasure of the Māori people. Historically it was an important source of food. Possums and rats are the two biggest threats to this bird.'

The fence is a thing of wonder, I am told – the mesh is fine enough to keep out most of the imported predators. Though the mice need extra attention, and there are small enclosures that are protected against even them. But what of the unwanted wildlife that was in there before? A massive exercise in the use of trapping and poison is what happened.

The statistics are quite dramatic. Starting in 1998, this was the very first attempt at a complete eradication of 13 species at the same time. The usual suspects were the targets: rodents, cats, possums, mustelids and rabbits. And for the first time both hares and hedgehogs were also included.

In 1999 50km of tracks were cut through the park for bait stations and trapping. In just 8 weeks more than 1,000 possums were killed – 3 tonnes of them. While possums are notable for preying on birds and their eggs, they also have an impact on the vegetation – their browsing is estimated to have removed 400 tonnes of plant matter a year from the valley before the fence was erected. This level of attention was having a serious impact on regionally rare species such as the northern rātā, tree fuchsia and kohekohe.

Next step, and all of this was undertaken with correct licences in place, was the distribution of cereal baits containing brodifacoum by air and by hand. Over two days 4.5 tonnes of the poison was spread around the reserve.

Bait stations and tracking tunnels were installed to monitor wildlife, and surveys for larger mammals undertaken too. The last four possums were removed in November and the last hedgehog – named Custer – trapped in December 1999.

The following year the last rabbit was shot. Since then only minor incursions have been noticed and dealt with, meaning that this was the first mainland successful eradication in New Zealand. Obviously, it is easier to do on real islands.

This is an 'island' clearance in all but name – how feasible is it to aim for the moon? For the entire country to be treated like this?

There is another question that remains largely unaddressed by this programme and the nationwide action of Predator Free 2050, though Alick Simmons certainly alluded to it. The way we treat an animal changes with the name we give it - and as soon as we call it a pest, humane treatment is often thrown out of the window. During the lockdown of 2020 I was being interviewed by Chris Packham and Megan McCubbin for the *Self Isolating Bird Club* – a quite wonderful positive to come out of the pandemic, a small scale, low-budget version of *Springwatch* – and I have this distinct memory of coming to the end of an answer and using the word 'pest', and how that caused a chill over the internet.

The problem with giving labels to animals is that it very swiftly 'others' them, reducing the need for them to be treated as individuals with the capacity to suffer and to fear. This makes it easier to do things to them that would not be done to other members of the same species in different contexts.

The New Zealand case has raised a different, and somewhat alarming, perspective picked up on by Michael Morris in 'Primary school education resources on conservation in New Zealand over-emphasise killing of non-native mammals', a paper published in the *Australian Journal of Environmental Education*. He notes the contrast between the government-approved National Curriculum for education in New Zealand, which 'emphasise[s] values of empathy and respect for all life,' and the educational resources that promote 'the view that non-native mammals should be killed. Some resources go further in giving instructions to children on how to do this, and how to source kill traps'.

The concern he notes is that of the 'increasing evidence that performing or witnessing animal abuse is a causal factor for future violence towards human and non-human animals'.

I do find this interesting and still need to think about it. I can only imagine, at the extreme end, that working in a slaughterhouse would tend to reduce your empathy towards the animals you killed, as if not, surely you would go mad.

There is a real sense that the wildlife-management world has become dominated by the New Zealand way of thinking, and these conservationists have become the go-to experts when other countries try to tackle similar issues. But just because it is a dominant way of thinking does not mean it is the only perspective.

I was doing some lectures in Oxford and when the academic who had booked me found out about my latest fascination he insisted I get in touch with an old friend of his from New Zealand, Professor Wayne Linklater.

There followed a rather cack-handed failure to understand time zones properly on my part, leaving me deeply confused when a Zoom meeting was booked for what would have been 4.30 a.m. in New Zealand, making me wonder what sort of a man this Wayne was. Turns out he was the sort of man who now had a job at Sacramento State University, California.

Our conversation left my head reeling – he was suggesting the most outrageous alternative views in the most gentle of manners. The good-natured iconoclast also added to the pressure by saying how much he wanted to read this book ... do I agree with all of his ideas? Well, sort of, but then again, no – look, here is what I learned and let's see how we feel at the end of it.

I did not notice it at first, but there was a hint as we started to talk. I was explaining how I wanted this book to be one that encouraged people to take ecology seriously, to accept that this is a deeply complex field of study. 'It is not just members of the public who don't treat ecology seriously', Wayne began, 'it is our peers in the conservation world too, who fail to appreciate the complexity of the problems we face and tend to want to offer simple solutions.'

If there is one thing we both instantly agreed on, it was that simple solutions offered to complex situations tend to be false, and those offering them either foolish or duplicitous.

'We lack humility in the face of ecology,' Wayne continued. 'We can't fight nature. Well, we might win the odd battle here and there, but we will never win the war. We also need the humility to recognise that we are entering into this world with our own set of values personal to us. We develop our own philosophy over time; that colours what we do, from what we choose to study to the choices of action we tend to take when confronted with a conflict. These are subjective. Science needs objectivity, so we need the humility to recognise the baggage we bring.'

I did not realise that in other disciplines published academic papers can come with what Wayne called a 'reflexivity' – a moment to explain why certain choices were made – offering insights into the fundamental values that guide choices at all stages of work, from inception to communication. I explained that this was why I was going to be talking about my 'prejudices' at the beginning of the book – I want people to know that I am aware of my own values – and that I recognise they colour my thinking.

'If we look at the sorts of wildlife conflicts we face around the world through the lens of climate change, I think it is useful,' Wayne said. 'For decades the climate debate has been about what we should do – whether we should adapt or mitigate – because we can't turn back time and not burn the oil and coal. And so with wildlife that we have introduced to new lands. We can't undo it, so we need to either mitigate – by which I mean, control by trapping and killing – or we adapt.'

Wayne is interested in the idea of what we are trying to achieve by setting out to eradicate a species – what are we trying to recreate? Is there an ideal place and time where everything was 'right'? Clearly not – everywhere we humans have gone, we have created change. How can we say which bit of this utterly changed planet is the perfect model for us to strive to achieve?

'I think one of the worst aspects of this way of thinking', Wayne continued, 'is that it requires that we identify an enemy. Targeting that enemy is not only futile, but it brings out a really nasty side to the conservation world. Once a species has had its name changed – to pest or vermin or alien – it is much easier to get away with doing things to it that would not normally be considered reasonable.

'At the heart of my thinking', he said, 'is that the ends do NOT justify the means. A lot of conservationists are, I am afraid, like developers and oil miners – they believe they have the power to overcome nature. We need to look at the big picture, we need to learn how to live with nature, not dominate it. And as I have already said, that will take some humility on our part. We are already past the point of no return. So now we need to engage with the ecological and social realities, rather than stay in a fantasy world.'

So far, so theoretical, but I was keen to find out what he actually meant, practically, by these assertions. How did this apply to his old home of New Zealand, and in particular Predator Free 2050?

'In 2018 I wrote a paper for the journal *Conservation Letters*. It might not have made me very popular with the organisers of the project, but I think it is a valuable critique.'

I had already got the piece open on my computer – there is certainly a 'no holds barred' approach that was more combative than he seemed as we talked. In it he writes, 'The goal is unachievable with current or near-future technologies and resources. Its effects on ecosystems and 26 other mammalian predators and herbivores will be complex. Some negative outcomes are likely. Predators are not always and everywhere the largest impact on biodiversity.'

'I was pleased with the positive response I got from many people after this was published,' he said. 'The mainstream view is so dominant that those who diverge from it can often feel insecure about speaking out. I was really keen to raise the key differences between mitigation and adaptation. Mitigation is what Predator Free are doing – they are using force to try and create a change. I believe this will not work. And the process of undertaking such an enterprise sucks the oxygen from the alternatives.'

Okay, but, still, what would he do?

'We know what we need to do – and we have the means – we need to create sanctuaries for vulnerable species and look to managing the wider landscape for the benefit of biodiversity.'

So, this would mean leaving the five target species – apart from within the sanctuaries – to just get on with it, and wreak havoc unchecked? I am bothered by this. Extinctions will undoubtedly follow this course of action.

'Okay, first, I think a lot of people suffer from a *Lord of the Rings*-induced misapprehension about New Zealand. Yes, it is a truly beautiful country, but it is also a deeply changed country – industrial agriculture has wrought so much damage. The reason that native wildlife is in such a state, teetering on the edge of extinction, is in large part down to what we have done to the land. Focusing the attention of us all on the impact that the five species of the Predator Free plan have had is a brilliant piece of diversion. All the energy for conservation, all of the money and social capital too, is being poured into this one pot. Not only do we not have money for other approaches, but also by burning up the political capital, we will find it harder to get more money when Predator Free fails.

'By adding in flag waving', Wayne continued, 'we end up creating a scenario where great suffering is sanctioned as a nationalistic endeavour. We should be better than that. This is really just a cover for the destruction that agriculture and industry have already done to New Zealand.'

He is clearly angry, even now in Sacramento. And he was far from finished.

'There is a move to encourage members of the public to take part in backyard trapping – to get community groups involved. This is so depressing. Again, it is the loss of social capital that will result from the failure of the project. Will we be so easily able to get people onside again for a nationwide conservation project? I doubt it.'

It was beginning to feel like Wayne was taking a very polarised view. And that bothered me. As much as I love some of my animal rights-supporting friends, I find their intransigence difficult to take. If we are not allowed to intervene, then we will have extinctions. Well, it turned out that Wayne was not completely opposed to killing – in fact it seemed that we have quite a degree of common ground when it comes to the seemingly intractable binaries we are increasingly encouraged to form. I have always felt that ecology necessitated a grey area, that there had to be room for things not to be so clear-cut. And this is just what Wayne went on to explain. He would kill predators to make the sanctuaries work. But better still is the move to adaptation.

'You realise that in 200 years, the rats that are now on New Zealand will have evolved enough differences with their starting population that they will become their own species, and there will probably be a move to protect them. Ecology adapts. If we have landscapes that are rich and diverse enough, the native species will mostly find a way to survive. But as I have already said, we need to take a step back and look at what we are trying to achieve.'

Well that is obvious, I thought: trying to stop species, found nowhere else on the planet, from going extinct.

'You think of the loss of a species is calamitous – but that is only because of your personal values. The ecosystem is still going to continue sequestering carbon, cleaning water and generating oxygen. Decay pathways will work, as will nutrient cycling. We have decided that the species is the pinnacle of nature, yet obviously it is not. They come and they go, and we give them value and try to protect them. But nature continues. Look – Antarctica is warming up, it won't be that long before plants start to grow there. Are we going to eradicate them because they are new? No, we are going to just accept that nature is changing and adapt. Of course, we are accelerating that change. But if global society was serious about the problems, we would pull back from this pathway to destruction.'

I remember reading a very provocative book by Fred Pearce, *The New Wild: Why Invasive Species Will Be Nature's Salvation*. In this he argues that we should take a step back from trying to control every incoming plant and animal and consider whether they will contribute to the health of the wider ecosystem.

Pearce argues that we are stuck within an orthodoxy of green xenophobia, and that clouds our judgement. Though he is not calling for a free-for-all – he recognises that some invasive species do cause havoc, but that is the measure, not that they are newly arrived. So there are possibly different ways we could look at the problems our inadvertent dispersal has caused.

But it seems like I am dancing around the crucial element here – so I had to be direct, and ask Wayne whether he really valued the concept of saving species?

'We need to be prepared to lose some species, and accept that others will only survive in zoos or highly managed settings. You know, this does not make me popular at home, but I do not associate the loss of kakapos, for example, with a large calamity. What I consider a large calamity is the loss of ecosystem function. We cannot go back – there is no point where everything was okay that we can reach. We have to look forwards. And we have to consider what we need to survive.'

The use of 'we' interested me and I had to ask for clarification, and yes, Wayne was referring to us – *Homo sapiens*. This shocked me. For so long I have been among people who are working with complete dedication towards the benefit of other species. It is almost like we have forgotten about our own species.

'Conservationists care about other species,' Wayne continued. 'Most of the rest of the world do not hold them in such high regard. They are bothered about having food to eat, somewhere to live, clean water to drink and a government that is not going to arbitrarily decide to kill their children. If conservation is going to win, we need to bring the large majority of the world with us.'

Wayne has been on a journey from conservationist to environmentalist – from species to systems. And along the way he has realised that one very crucial thing is often left out of this debate, and that is justice. Jingoistic, nationalistic flag-waving over particular species is getting in the way of tackling a global systemic decline. To summarise that Wayne said simply, 'No justice, no peace.'

'It is not that biodiversity has no intrinsic value', he said, 'it is just that we need to weigh up the values of a species versus the values of clean air, for example.'

We said our goodbyes, but I was left somewhat reeling from Wayne's assault on much of what I took for granted. His point of view may not be accepted in conservation circles, but I think we are foolish to ignore this very different perspective. And possibly the most important perspective shift is to embrace a world view that is less geared towards killing for conservation.

Wildcats

We have already looked at the 'when' – to what point in time are we trying to return. And as an idea, it is most certainly found wanting. There is no prelapsarian idyll to which we can return the world. Species are always on the move, the biosphere has never been static. And if we discount anthropogenic dispersal, there is also the dispersal we are generating through climate change. Animals and plants move to cooler places until they run out of land.

So how about we ignore the 'when' and focus on the 'what.' What are the species we are trying to save, and in particular, how pure do we want them to be? For some species, the great threat comes as much from a lack of purity as it does from hunting, habitat loss or climate change.

When I was at school I remember being taught that the definition of a species was pretty much like the early definition of an atom – something 'uncuttable'. A species was identified as one that could only breed successfully with others of its own kind. This is the 'biological species concept'. This is obvious with, say, a hedgehog and an elephant – there are numerous impediments to success. But a horse and a donkey? There, the impediments are far fewer and you famously get an infertile mule. Similarly there are ligers and tigons in captivity – though the chances of these happening outside the bars has significant geographic obstacles – and they too are have reduced fertility.

Species were identified, initially, by morphology – by how they looked and their inner anatomy. More recently there has been a move towards using genetics, which has thrown up some interesting relationships. For example, we now know that fungi are more closely related to people than they are to plants – though there is unlikely to be any chance of successful mating in any of those cases.

The problem for conservationists is when we get to the very small difference between species – there are competing views. For example, we have the concept of 'twitcher-driven speciation', to which I was introduced by the sparrow expert, Denis Summers-Smith. He wrote, 'Fifteen species of sparrow were listed when I started my sparrow studies in 1948; this has now grown to 26 species. Some years ago I wrote to Stephen Jay Gould, the renowned palaeontologist, evolutionary biologist and popular science essayist, suggesting that this rapid evolution of new sparrow species lent support to his hypothesis of "punctuated equilibrium", the idea that the evolution of new species occurs in periodic bursts separated by long periods of stasis. He dismissed my proposal with contempt, attributing the increase to no more than "twitcher-driven speciation", a cynical attitude taken by the authors and publishers of new checklists of the birds of the world to stimulate their sales by giving birdwatchers more opportunities to increase their "life lists"!'

How pure those species are is the question here. And the best case study comes in the form of the wonderful Scottish wildcat. It is somewhat shocking to realise that the most endangered cat is not the tiger or the snow leopard, and it does not live in exotic places – unless you consider the Highlands of Scotland to be exotic, in which case it definitely does.

I have had the good fortune to meet a few wildcats – in captivity – and while at first glance they might be mistaken for a beefed-up tabby, that moment does not last very long. These are cats with attitude. They don't exactly exude menace, more a sort of confidence that leaves you in no doubt as to their capacity. If a wildcat was a human it would be a young Marlon Brando – leaning, looking moody and beautiful.

The Scottish wildcat has not always been Scottish. It has only become so by default, as we have exterminated this once-widespread animal across the rest of Britain. It was part of the postglacial carnivorous fauna, and alongside the brown bears, wolves and lynx, the wildcat helped maintain a landscape of fear for herbivores.

This was a time when Mesolithic humans moved back onto the ice-cleansed earth, around 11,000 years ago, and joined in with the creation of the landscape of fear. And as those humans began to assert dominance, so the carnivores began to retreat. We are really not very good at sharing. The lynx were gone by around 800CE, brown bears clung on until around 1000CE, and wolves were finally killed off by the late eighteenth century. But the diminutive wildcat kept its claws in the land.

Gamekeepers, farmers, cars, all have contributed to the decline of this magnificent predator, to the point that some consider it functionally extinct in the wild. This would mean that the only wildcats left out there are too dispersed to form a viable population, and are at risk of the greatest threat – the lure of a farmyard moggy.

This is when the definition of what a species is becomes so important. When wild and feral cats mate, they produce fertile, hybrid, young. These youngsters then grow into adults who follow their natural urges and in turn create additional and different mixtures of genes – all of which leaves the landscape with cats that are not really wildcats.

Extinction by dilution is an undramatic affair. And it is also very hard to determine. Are there are any pure wildcats left? How do we determine what a true wildcat is? What should happen to those that are considered impure? Are they considered a threat to the continued existence of this magnificent predator? Are they to be killed?

Death is the fate that will befall any hybrids that get caught, as conservationists attempt to protect the remaining purebloods. Working out which is what is tricky; there is a technique that uses the pelage – basically what the coat looks like. They have solid black and brown stripes. Spots, broken stripes or white fur are all indications of hybridisation. And their tail is thick with perfect rings, and no dorsal stripe running off the spine. Again, hybrids tend to have broken rings and dorsal markings. And then there are genetic techniques that look at the purity of the individual. Both methods of determination are flawed – if the type specimen in the genetic study was not pure, this could lead to pure wildcats being killed as they don't match.

Does it matter if the waters are a bit murky? That might sound sacrilegious, but it is an important question. After all, despite it seeming obvious that a species is a clearly definable 'thing', it is actually a concept that we have created to try and pop life into neat boxes. In reality, species are not, or at least are often not, quite so conveniently contained.

So what is it we are trying to protect? And are we trying to dip life in aspic, to keep species the same? In the name of preserving biodiversity this is, pretty much, what is happening. Not aspic but deep freeze – cryopreservation – living tissues and cells are being stored.

The Molecular Collections Facility at London's Natural History Museum, is not on the usual tour of the beautiful building in Kensington. In fact it is in Harwell, Oxfordshire. Here they have the capacity to store more than two million tubes, each containing a different sample from a different organism. The freezers keep them at -80°C (our home freezers only get to -20°). There is also the capacity for 60,000 specimens to be kept in liquid nitrogen at -185°C.

The sobering thought is that some species now only exist within these tubes.

Are species objectively or subjectively valuable? Certainly Wayne has raised an alternative view. Are they valuable because we subjectively give them value, some species more than others; or are they objectively valuable – a value that is present whether we are there or not?

Or should their value be measured in the manner with which they help maintain an ecosystem?

The wildcats do not want to mate with feral cats – they much prefer their own kind. And this is why the work of breeders like Derek Gow – who does more than water voles and beavers – is so important. I do not know whether 'his' wildcats are 100 per cent pure, and it is pretty clear that that idea is a nonsense anyway. But when you are in an enclosure with them, they 'feel' pretty wild. The more of these there are in the wild, the greater the chance that mating will take place between them, and not with the feral cats.

And there is no reason why these cats should be the preserve of Scotland. Wouldn't it be wonderful to one day catch a glimpse of a wildcat near where you live?

CHAPTER 10

Compassionate conservation

Australia, unsurprisingly, is also beset by the challenges of species being where they might not be best placed. Isolated 50 million years ago, life in Australia evolved into an amazing and diverse array. I have never been, but the variety of life, so different to that with which I am familiar, makes the prospect rather exciting. Australia is home to the mala, the spotted quoll, the western barred bandicoot, and a small wallaby who was named the woylie. Though it is possibly the greater bilby that wins – a long pointy snout like an oversized shrew, hare-like ears, black and white tail, and strong forelimbs to dig burrows up to 3m long. Their digging means they are ecosystem engineers, turning over 20 tonnes of topsoil a year in their quest for food and shelter.

It is easy to imagine an untroubled life for these fluffballs. Then humans arrived around 70,000 years ago and settled in, and managed about 20,000 years of relative harmony before a combination of improved hunting, fire use and climate change set about extirpating the megafauna. This left the continent far more open to other species stepping in to fill vacant niches. Isolation was on its side until the arrival, more than 5,000 years ago, of dingoes. Now very much part of the fauna, they did have a dramatic impact.

But it was not until 1788 that the real trouble for wildlife (and the Indigenous people) began, as European settlers settled. Ignorant as they were of the relationships already forged with nature, they declared the land empty and up for grabs. And along with the taking, came the bringing – of new species. Australia has 56 introduced vertebrates, falling into three categories: invasive, pest or feral.

Feral goats, pigs, donkeys and camels are all culled in Australia, in an attempt to control their impact on vegetation.

Much of the energy put into this is due to their nature as a competitor to livestock. But they are also considered threats to the value of the ecosystem, as they degrade the flora and increase the risk of desertification.

Rabbits have long been in the sights of land managers, and have been attacked in many ways with varying degrees of success. You may remember a film called *Rabbit-Proof Fence*, powerfully telling the story of the inhumane child-removal policy of Australia, where Aboriginal families were split up by the white settlers. The story follows three girls as they flee the abuse of a European-run camp and try to return home. They use the fence that was built to try and keep rabbits out of agricultural land as their guide as they walk 2,400km over nine weeks.

Fences were only a part of the strategy; more devastating has been the development and release of diseases aimed at the rabbits – most famously myxomatosis, which became the first intentionally introduced virus used for pest control when it was released in Australia. The virus has a natural home in the Americas, where it causes only mild in the cottontail rabbits native to those areas, but in European rabbits, who evolved in Asia, it was found to be highly 'effective' at causing death. A miserable and cruel death, if you have ever had the misfortune to seeing a rabbit in the final stages. In 1950 it was released in Australia and saw a 99.8 per cent death rate – though over time resistance developed. So in 1995 rabbit haemorrhagic disease virus was released to further the cause.

While this is at first considered to be 'pest' control, conservationists also recognise the damage rabbits cause to arid-zone ecosystems. And some have started to complain about the attention paid to predators and the lack of effort put into working on herbivores.

But it is to the carnivores we must now attend, because their impact is nothing short of dramatic. When Europeans first arrived there were 273 endemic terrestrial mammals living in Australia, 28 of whom have now been wiped off the planet. Fifty-six more are threatened with extinction and 52 species are listed as Near Threatened, according to the IUCN Red List criteria.

The principal culprits are foxes and cats. Be prepared for some more alarming numbers. Each year it is estimated that feral cats kill around 1.4 billion native animals in Australia. That is around the number thought to have been destroyed in the devastating bushfires of 2019–2020. On top of this, pet cats are also responsible for the deaths of a further 390 million animals a year.

While cats came as pets and pest-control agents, foxes arrived in the 1830s to enable the 'noble' tradition of fox hunting to continue in this new world – with predictable (thank you hindsight) consequences. Their spread corresponds to the decline of many medium-sized mammals, including brush-tailed, burrowing and rufous bettongs, numbats, quokkas and bridled nailtail wallabies. These animals find sanctuary on islands and in areas where foxes are absent.

Interestingly some people have posited a counter to this, arguing that foxes actually keep numbers of rats, rabbits and feral cats down and that by removing the foxes you may end up causing an explosion in these animals' numbers.

To conserve, in particular, Australia's native fauna, it seems obvious that the predators must go. But how? Traps, shooting and poison are all used to control foxes, but there is no chance this will result in eradication, as that is simply too big a problem. This means that foxes will be killed in perpetuity in the name of conservation. The most common technique is poison bait.

One project is called Western Shield. Since 1996 this conservation effort has been broadcasting 1080 poison over 3.8 million hectares of Western Australia. This poison was chosen as it turns out that its active constituent of sodium fluoroacetate, while being exceedingly toxic to foxes and cats, dogs and humans, is found naturally in some Australian plants, leading to native wildlife gaining tolerance to its effects. And as we have already seen, it is the poison of choice in New Zealand.

The 1080 poison is injected into salami-like sausages, called Probait®, which are then dried to make them hard and less palatable to native animals, although they are attractive to foxes. Cats are very sensitive to 1080 but prefer live prey, so they do not normally eat the dried meat baits used to control foxes. Parks and Wildlife Service staff have developed smaller, tastier and moister 1080 sausage baits, more appetising to feral cats, called Eradicat®. 1080 breaks down quickly in the soil without any environmental side effects. However, baits, and the flesh of animals that have died from 1080 poisoning, can remain toxic to dogs and cats for months.

I thought it might be instructive to look into how you could tell whether your pet had eaten any of this bait, and the results I found from Crookwell Veterinary Hospital in New South Wales, well versed at dealing with cases, gave alarming reading:

> *Clinical signs of 1080 poisoning are usually noticed within half an hour of ingestion, although symptoms can take more than 6 hours to manifest. Initial symptoms include vomiting, anxiety and shaking. These quickly develop into frenzied behaviour with running and screaming fits, uncontrolled paddling and seizures, followed by total collapse and death from lack of oxygen to the brain. Rigor mortis sets in quickly.*

With the best will in the world, it is hard to read that set of symptoms and think that what we are doing to wildlife in Australia through no fault of its own, could in any way be described as compassionate.

While it was still legal to kill moles in the United Kingdom with strychnine, I interviewed an ecologist who explained why they were the only animals on which it was still used. And that is because the moles die underground, so we don't see the agonising contortions that can be so severe that they break the animal's spine before death. Poisons allow us to remain ignorant, but that is a dishonest approach, leaving the 'spreadsheet' lacking.

Maps provide clear warnings of where this poison is scattered, and there are signs on the ground too, so dog walkers know where to avoid. There is a lot of poison – Australia's Parks and Wildlife Service distribute around a million baits per year, dropped every three months from aircraft.

Clearly, to have been going on for more than 20 years this must be an effective campaign? Well, it turns out it was for a while, and resulted in the wonderful woylie, along with the tammar wallaby and the quenda being removed from the list of threatened species. Unfortunately, woylie numbers dropped again and they have been re-listed. As is often the case, while the invasive species might be the obvious problem, there may well be many other factors at work.

There are more than 3.8 million pet and 2.8 million feral cats in Australia. As we have seen, they have a staggering impact on native wildlife. But cats have a very effective protection racket going on with people, somehow convincing us that they are to be loved with every ounce of our being. So even my lovely, environmentally conscious neighbours turn a blind eye to the corpses their beloved pussies accumulate.

There have been efforts to reduce their impact. Newhaven Wildlife Sanctuary in central Australia's Great Sandy Desert, has managed to remove the cats from the 2,600km^2 behind the cat-proof fence. And pet cats are now subject to serious curfew and containment laws across much of the country, with most states seizing roaming cats.

There has been, of course, pushback from animal lovers, with some arguing that feral cats should be considered native fauna and accepted into the fold. While that might seem absurd, there is also the fact that there is really no way to eradicate them, so you are either confronted with a perpetual task of control and the maintenance of cat-proof areas, or just working out how to get along with them.

Though there has been some work done that might just hint at alternatives. European settlers in Australia have fought a long war against the dingo. It was a predator of the sheep that were being grown on what was once others' land. But according to the work of ecologist Arian Wallach, 'Across the continent, the presence of dingoes is a major predictor of low fox and cat densities and high survival of endemic small mammals.'

This means that an unintended consequence of killing dingoes has been the death and potential extinction of native species. What other lateral thinking is out there? For example I love the solution offered to the problem of little penguins (the species, not their size) being killed by foxes on Australia's Middle Island. Numbers were down from 600 to just 10 over five years of fox predation. Attempts to eradicate the foxes with den fumigation, trapping, poison and shooting all failed because at low tide more foxes would recolonise. So in 2006 a trial was started using Maremma sheepdogs, trained to guard the colony. Since then, fox predation has stopped.

Perhaps the most bizarre killing in Australia is that of the most identifiable animals of the country – the kangaroos. In the name of conserving landscape from desertification and all the loss that will cause, kangaroos are killed. Though this is not something you will see much about in the tourist brochures. In fact I was only made aware of this from the very determined work of a friend of mine, Jenny. I met her dancing and was always impressed by her playful passion. She invited me to a demonstration she was organising outside the Australian High Commission in London.

The figures are startling – a quota of 236,350 in 2023. Killed in the name of conservation, but actually, according to Jenny and many others, to feed an industry for meat and leather. The real reason there is such desertification, she states, is because of the massive overgrazing of livestock. 'The scale of this', she said, while handing out leaflets to the commuters passing by, drawn to the graphic images the campaigners had on display, 'is something that most people are not aware of. So many kangaroos killed – and they say it is for conservation, but they are lying. It is for money – it is driven by greed and is obscenely cruel.'

Everywhere I go it seems that there is killing, and if I am being honest, I am getting tired of the lack of imagination. I feel there must be a better way. One solution was mentioned, sort of, by Chris Jones while discussing the future of wildlife management in New Zealand. He talked of genome mining to find something that could affect just the one target species. So I looked into it and found that there really is a technology that could step in to help control some wildlife without the messy business of killing, by reducing the fertility of the target population. But while it might seem obvious that I would be excited, there is a problem … it is a very new and very untested form of genetic engineering.

I have a bit of history with this world. In 1998 I was recruited to edit a small magazine, called *Splice*. It was the journal of The Genetics Forum and was a really interesting opportunity to critique the burgeoning biotech industry. In particular there was a lot of attention paid to the rise of genetically engineered crops. The Soil Association also got me working on a report for them, called *Seeds of Doubt*, which involved me travelling around the US, meeting organic farmers who had had their livelihoods destroyed by the spillover of genetically modified crops from neighbouring farms.

So, I was definitely in the sceptical camp, but my main concern was less about the technology itself, more about the way it tied farmers into a system of agriculture that relied entirely on external inputs of agrochemicals.

With this baggage in mind I approached CRISPR with caution. CRISPR is an acronym, but the full name, 'Clustered Regularly Interspaced Short Palindromic Repeats', leaves me none the wiser. This led me to turn to Dr Ricarda Steinbrecher – founder of the public-interest research organisation, EcoNexus, and an old friend. By way of transparency, I am also a director of EcoNexus – but in a not very active form.

I first met Ricarda through the peace movement. Before she moved to the United Kingdom she had worked in Germany with one of my all-time heroes, Petra Kelly, a founder of the German Green Party. A genetic scientist by training, Ricarda now spends a lot of her time working with the UN Convention on Biological Diversity (CBD), supporting the work of people and communities who do not have as much capacity to engage. In particular, as part of the CBD International Expert Group she is looking at the risks of new technologies and if they could have an impact on communities or ecosystems.

We met over a coffee in Oxford's best bookshop, Blackwell's, to get an introduction into what this new form of genetic engineering is all about.

'CRISPR is a tool,' she explained. 'A clever device, based on a bacterial antiviral defence system – it is guided by synthetic RNA to a point along the cell's genome where it then cuts the DNA. The cell will try and repair it and in this case gets tricked into inserting the desired gene. And if this automatically happens in each offspring, then you could drive a gene or trait rapidly into a population. Well, at least in theory, that is.'

I got the feeling that there was vast amount of detail she was leaving out, and for that I thank her. The reason this has got people in the conservation world excited is that, theoretically, it could be used to cause the spread of a characteristic through a population that, for example, leads to the disappearance of that population or potentially even an entire species. So, you could infect the mice on Gough Island and make them sterile, or to only have male young, by removing the X-chromosome … a technique called the X-shredder.

'When you were first getting involved with genetically engineered food', she continued, 'do you remember how one of the central problems was a lack of risk assessment, of regulatory framework, into which research could be undertaken? It is the same again here, and through the CBD we have been pushing for specially adapted risk assessment for Gene Drive Organisms, and for the strict application of the precautionary principle. The risks for biodiversity are extremely high, and you cannot just do field trials to test the performance of these organisms, as once you release them, they can go wherever they want, be it insects or mice or rabbits. Knock-on effects can be devastating for ecosystems and biodiversity. The trouble is there is no agency for compliance, and we need international compliance.

'There are a lot of people very excited by the potential of this technology,' she said. 'And this has brought a great deal of money into the arena, which amplifies the voices of those who are developing ideas, in turn making it harder for our voices of concern to be heard.'

The reason for caution has been around for a while. In 2017 there was a paper published in the *Proceedings of the Royal Society* which stated, 'Self-replicating gene drives that can spread deleterious alleles through animal populations have been promoted as a much needed but controversial "silver bullet" for controlling invasive alien species.'

Ricarda and her team have done a thorough job of looking at which species are being targeted and which technologies are being developed, and which are the labs involved. And while it is, as one would expect, insects that get the most attention – thanks to the damage they cause as agricultural pests and spreaders of disease – increasing attention has been paid towards mammals.

But this is still a long way from ever being let loose among vertebrates, though trials with mosquitoes and agricultural insect pests are well underway in the laboratory.

Is there really an option for compassionate conservation? The phrase 'Compassionate Conservation' got public airing in Oxford in 2010, at a conference hosted by the wildlife charity, the Born Free Foundation. Yet another moment of hindsight – would it not have been useful for me to go to that? Though at the time I was in the middle of writing about other things and it passed me by.

I feel very drawn to those principles. I instinctively feel opposed to killing. I have had to do it – crushing the head of a beautiful artic tern under my heel when I found it with one wing almost completely cut off from a collision with an electric line. But I hate the thought of it, so much so that I have not eaten meat for more than 35 years. I would love there to be a way of fixing the messes we have made that did not rely on such lethal action.

Yet I am also realistic enough to realise that there are times when inaction is not the absence of violence that it is naïvely supposed to be – but just slow violence. We may not be setting the poison but we pulled the metaphorical trigger when we introduced the rodents to an island; we are still responsible – unless we are willing to embrace Wayne Linklater's demotion of the species as a target for conservation.

A key mover in this world is Marc Bekoff – who was, of course, at the 2010 meeting. He has written extensively alongside an academic career, and has edited many of the ideas that surround this in a book called *Ignoring Nature No More: The Case for Compassionate Conservation*. It includes masses of brilliant ideas, focusing mainly on the philosophical rather than the practical – and it is one of my most heavily annotated books. I have a secret code of pencil marks – which only I can determine on a re-visit … sometimes … other times I am at a loss as to what that squiggle actually meant.

Compassion – it is a good word, coming from the Latin 'to suffer with.' It requires an exercise in empathy.

When you look into the aims of the compassionate conservation movement you find they have managed to distil this down to four overarching tenets. First, and who can deny this, is 'Do No Harm.' This is not just something that looks good on a t-shirt. It is thought to have been coined as an idea by Ancient Greek physician, Hippocrates, and is at the heart of medical bioethics to this day. The instinct to intervene should be carefully scrutinised and not always followed. It may be better not doing something – or even doing nothing, than to risk causing more harm than good.

Second is the idea that 'Individuals Matter.' This is quite challenging for me. Yes, I am very conscious that every individual matters, but I am recognising that I have a utilitarian heart (though there may be some who would suggest that is an oxymoron). 'Individuals Matter' acknowledges the intrinsic value of individual members of a population and asks that we resist the tendency to reduce them to just being part of a collective.

The third tenet, 'Inclusivity', encourages us to acknowledge the value of all wildlife – irrespective of sentience, of population size, and of utility to humans. The final precept is that of 'Peaceful Coexistence,' which again is challenging in a number of the cases I have investigated. The compassionate reaction in conservation is for us to think less about how we imagine the world ought to be, and more about the manner in which we ought to engage with the world as it is. In a paper he co-wrote in 2018 Marc says, 'It demands that the first instinct in conflict situations should be to critically examine and in many cases modify our own practices, rather than pursuing acts of aggression against wildlife individuals.'

Marc was clearly someone I needed to talk with, and while I would have loved to join him in his home beside the mountains in Boulder, Colorado, I had to rely on Zoom again.

For someone more than 20 years my senior, he is remarkably spry – many decades as a vegan, a one-time elite athlete and, it turns out, the possessor of a brilliantly sharp mind.

To be honest I was somewhat intimidated. I am averse to conflict (so why on earth am I writing this book?) and imagined that he would find my rather vague-an stance on matters around killing to be worthy of attack.

From the outset, he was completely uncompromising. 'I want killing off the table,' he said. 'When killing is a possibility in a conservation setting, it is almost invariably the choice that is taken. We think we have the right to intervene – but human exceptionalism is not something I want to be part of. And if you take the time to look at our interventions, we have not been doing a great job of it.'

Uncompromising, yes, but at no point aggressive. 'I like to think I'm a nice guy,' he said with a smile. 'I'm not up in people's faces being argumentative – and this is why I get invited to the table to talk over ideas. I just want to raise the alternative perspective, the one that does not involve killing. And I want them to realise that they are making an active choice to kill – they are choosing that one species has more of a right to life than another.

'When I do talks I will often get a gasp from an audience when I ask them to imagine someone knocking on their door and saying that there are too many golden retrievers in the environment, so we have decided that yours should be culled,' he said. 'Yet the audience may be convinced that there are too many wolves, and therefore they should be shot. What is the difference? The wolf and the retriever will both suffer the same pain, their families will suffer the same loss.'

But his proposed lack of killing does not necessarily result in less death. And it is a death that we have caused, albeit through misadventure when rodents get onto an island, for example.

'A lot of people in this world are driven by very short-term thinking,' he said. 'They are looking at getting results from research that will lead to a paper being published or maybe getting tenure at a university. So when I say we should leave the non-human world to sort itself out, there are serious misgivings – because things won't sort themselves out in the time frame of a PhD, nor in a decade. In fact ecosystems will take centuries to reach a "conclusion" to the problems we have thrown at them. That is the timescale on which I am looking.'

This is a completely different perspective on these conflicts. I too have been driven by the idea that we must solve the problem now, but in many respects we are doing so for our own personal reasons.

'I find it strange', Marc continued, 'mostly we kill to make ourselves feel better, to feel like we have tried to clear up a mess of our making. Key to Compassionate Conservation is the idea that individuals matter. Every individual animal has an inherent desire to live, they have evolved over millennia adaptations to reduce their chances of being killed and increase the chances of having children – and I use that word deliberately. We all too easily dismiss animals through language – often not even consciously. Livestock? Really? These are pigs, cattle – and they do not give birth to commodities, they give birth to children. Given the opportunity these non-human animals will nurture their young; they will do the best they can to help them live.'

It feels like Marc does nothing to help my problem with the polarised mindset. I bring up the importance of the middle ground.

'You talk about this needing the grey areas to be considered,' he said. 'But whenever I see grey areas, they are always leaning towards killing. Often, if I don't bring up the idea of not killing, it will not even be on the agenda. I have been accused of having a very female mindset – as an insult. What nonsense. And I am also described as being radical, well, what is so radical about not killing animals? Describing someone as radical is a very effective tool used to marginalise one side of an argument.

'One of the bits of grey that I feel is so wrong is the discussion about the humaneness of killing,' he continued. 'I am not interested in killing animals "softly", I don't want them killed. And when you dig around into the literature and find out what some people pass as humane … so, one of the main poisons that are used are anticoagulants. These are described as humane in large part because we tend not to see the process – as the individuals who have eaten the laced bait slowly bleed to death. So how about a fast poison? Like cyanide? Well that is no good, we need them to suffer a slow death as it is good for us, our children and pets – because if any of us consume the poison, there is time to act. Then there is the third option – not killing them. How about having a conversation around that?'

I tentatively brought up the conflict surrounding trophy hunting – the argument that the money earned from killing a few of the charismatic megafauna will help fund the conservation of those that remain. Marc's response was very much in line with the rest of his thinking: absolute, but strangely non-confrontational. 'I am against trophy hunting,' he said, unsurprisingly. 'Conservationists play the numbers game. Look, if we kill this many, so many more will thrive. Going back to thought experiments with your dogs – how would you feel to be given the choice of having yours killed so that 40 others will be able to live happily? I use dogs because they bridge the empathy gap – they are a gateway species to deeper thinking. All it takes is for people to begin to understand that we have decided that some species, and some individuals within some species, have more right to life than others.

'It is now said we are in the Anthropocene, the age of humanity,' he continued. 'But really I think we are in a rage of inhumanity, obsessed with the idea of human exceptionalism, that we are in some way the fulcrum about which the world is revolving. We have not moved on much from the church that persecuted Galileo for daring to suggest that the world was not the centre of the universe.'

Despite his obviously strong feelings Marc never strays into unpleasantness – there is a lot that some supporters of animal rights could learn from him. 'I love a good peaty malt whisky,' he said, and here was a point of agreement. 'And when I've been at a meeting or a conference and we gather in a bar afterwards – especially over in the UK – and we share a drink together, however difficult the conversations might have been, when we start to spend time together as fellow humans, the common ground becomes apparent. We all share most of the same aims in life – this is where the connections start and change can grow. And maybe the recognition that despite apparent differences, we each matter. Just like the other individuals who live as different species.'

As warm as I feel to the arguments of Marc, I am left with a problem – his desire to take killing off the table will not take death off the table, it just means that we, should we choose that path, are a step removed from pulling any metaphorical triggers. Let us imagine some goats on an island. If left unmanaged – if not killed – they will strip the vegetation from the island, reducing a rich ecosystem to a pile of rocks, causing the extinction of plants and animals, including the goats. He argues that we should let time pass. But the act of not killing can cause more death and, probably, suffering than if killing had taken place.

Perhaps with a nod towards Marc and his school of thought comes a halfway solution. Dr Benjamin Allen from the University of Southern Queensland released dingoes onto Pelorus Island to see whether they could act as a biocontrol tool – eradicate the 300 or so goats with four sterilised predators. The experiment worked, and after two years there were thought to be only two male goats left. The removal of the goats led to a trophic cascade as the island bounced back into a healthy ecological shape.

This was not the first time such an exercise had been undertaken. Goats have become a problem on many islands where they were released to help feed lighthouse keepers or sustain shipwrecked sailors. In 1993, 16 dingoes were released onto the 70km^2 Townshend Island, where 3,000 goats were stripping the vegetation. It was brilliantly successful in that the goats were gone in two years. But … it took a further 15 years to remove the dingoes, who had moved onto eating ground-nesting birds.

So to ensure the dingoes on Pelorus do not cause such a kerfuffle, they had tracking collars, enabling the conservationists to find them and shoot them when their job was done. And just in case that did not work, they had a back-up plan – the dogs had a capsule of the poison 1080 implanted into their bodies, timed to go off after two years.

The process of researching that example took me on a journey from excitement to despair in a very short time.

It feels like there is no way that we can act and not cause harm – and yet inaction also causes harm. In 2022 a paper was published that went some way to ease my misgivings. Three different ways of thinking about these problems are not as entirely different as they might at first seem. Deontology, virtue ethics, and consequentialism are all ways in which we can judge whether an action is morally right or wrong.

Deontological (normative) thinking is at the heart of Marc's way of life. 'Do no harm' – essentially it is like a commandment: 'Thou shalt not kill.' It is a duty, a self-imposed rule, and in conservation this means that we should not intentionally harm animals.

A **virtue ethicist** would look at the problem from their own moral character – compassion is a virtue, and if we are to show good moral character we will display compassion to others, and by extension, towards wildlife too.

Consequentialism is not focused on the duty required by a deontological mindset, nor the good character of a virtue ethicist. The goal is the outcome – the consequences of an action or inaction.

Marc taking a position of 'do not kill' is admirable and I wish I had that certainty in my life. But it does make me want to pick at it, to unleash the 'what-ifs', like the goats on the island. This takes me back to the trolley problems near the start of the book. I came across a different version mentioned in a paper in the journal *Frontiers in Psychology*; it is not new but it very much appealed to my scab-picking mind. Thou shalt not kill. But … 51 miners are trapped underground – we can save 50 by killing one, we can save one by killing 50, or we can do nothing and allow all of them to die.

All the way through the writing of this book I have been struck by the impossibility of absolutes. I do not want to kill, but I will and have killed an animal in distress. As soon as the grey is accepted, we must become more willing to consider the messy, uncomfortable nature of it all.

Maintaining a good and virtuous position is, again, admirable. But this is very much not an ideal world. A virtuous person will, I would hope, euthanise an animal in distress. If I had stood by and watched the artic tern flap its nearly severed wing until it was exhausted and died of hypothermia – or was eaten by a passing cat – I would suggest that was the antithesis of a virtuous action.

Perhaps it is more than just theory behind Marc's ideas. I was lucky enough to have coffee with Prerna Singh Bindra on a chilly November morning in 2022. I had hoped we could find space inside the Vaults Cafe, in Oxford, but it proved too noisy, so we bundled up and sat outside. She is a formidable wildlife conservationist who has spent 15 years working in India before coming to Cambridge to do a PhD looking into the relocation of people out of protected areas. She explained to me the different attitudes she has noticed to wildlife, in particular the willingness of some – many, or in fact most – to consider killing an appropriate action. 'In India', she explained, 'we have culture and religion at the heart of what we do, and so we do not cull wildlife for conservation.'

She reminded me of the very different reality India has, compared to Oxford or Cambridge, for example. Every year around 500 people are killed by elephants and, of the 1.2 million snake bites that get reported, up to 40,000 people die. She is not naïve, she knows that wildlife is killed, often in retaliation. But there is no national policy of intervening with lethal force. The accommodation of wildlife is something from which we could all learn.

The point that Prerna makes is important: people have to be part of the equation, they need to believe in what is happening. Yes, we need experts, but we also need understanding that the situations of individual people may well be very different. For Monica Engel, a remarkable researcher from Brazil who has ended up in Newfoundland specialising in the human dimensions of wildlife and natural resource management, this is at the heart of her work. We booked a call early in 2022 and I had an illuminating hour in her virtual company.

'My job is to get a clear picture of what is unfolding as we talk about hunting or conservation,' she explained. 'I have to shed all my judgements, I have to put all the bits of Monica that "think to know" the answers in a box so that I can just listen. Observing like this does not mean I do not have an opinion, but when I am working I have to be neutral. You know, I started out as a biologist, but as I moved into conservation I realised that it is impossible to do that work without involving people.

'Take seal hunting,' she said. 'There are hunters who are there for business, there are conservationists wanting to ensure populations are stable, there are members of the community who will have a mix of views, and there are protesters campaigning against the killing. My job is to listen to everyone and try and find common ground. And it is fascinating – even with something as contentious as killing seals, there are many shared values and beliefs. You cannot judge the situation by looking at the extremes.'

There are practical problems to this approach. A young conservationist from Brazil in a remote community in Newfoundland – how do you go about gaining trust when feelings obviously have the potential to run very hot?

'Yes, it was very hard in particular to get access to the hunters; they are understandably defensive,' she explained. 'So I signed up to get my seal-hunting certification. I did all the training and this was my way into their community – I got to listen to them. Though I never actually went and did any hunting.

'One of the most fascinating things I found was that wildlife is often used as a scapegoat for social ills,' she said. 'It can be used to get at the government, scientists, outsiders – I think that this has always been the case to some extent but now, with social media, it is all the more visible. And it makes arguments become polarised. Fear and anger have been stoked; we are losing connection to community and nature, we are like crazy ants without their queen.'

While seals may be the obvious flashpoint, another issue that Monica is dealing with revolves around the management of reintroduced bison to the Yukon. This has gone well, but has not been without tension. These amazing animals do have the potential to cause damage, and they do generate fear. What I found particularly interesting was the analysis that showed how there was a 'social carrying capacity', which was lower than the more traditional measure of an ecological carrying capacity. The people would only tolerate a lower number of bison than the land could sustain. To help maintain bison there had to be a good relationship with the hunting community too – to conserve the balance that would satisfy ecological and social needs.

'I realised early on that all of these issues have to be addressed on the long term,' Monica said. 'This is never just a "date", you have to go into this realising this is a marriage.

'When I am out there, listening, I find it so sad that both the hunters and the animal rights activists enter into this debate with often so much hate.' And it is clear Monica feels this deeply. 'It drowns out moderate voices, it stops progress. I may disagree with certain actions, but as a scientist my personal opinion doesn't really matter. I try to document and understand all the views and put them into perspective; bring it to the bigger picture. I find myself asking people about their motivations for killing, or protecting a species from any form of hunting. I need to understand – we all do – if we are ever going to be able to find common ground. Because if we just shout at each other, we get nowhere. And many hunters want species to thrive.'

Another accommodation we are going to need to be showing in the years to come is for climate refugees. We already know that humans are going to need to move from parts of the world that become less agreeable to life, and so will be the case with wildlife. There are already small changes afoot – the blackcaps visiting for the winter in my garden are new; they used to spend winter in Spain. The greylag geese on Orkney are changing their behaviour with the climate. The RSPB believes that European breeding birds will move around 550km north by the end of the century.

The delicate scops owl, currently only seen on British shores as a vagrant, is expected to be breeding regularly here by 2075, when the magnificent hoopoes and the lovely yellow serins will also have become established in the new normal. So we need to be cautious about trying to hold the line in defence of an artificial image of what should be here – compassion and accommodation, and maybe a little thought, will be needed as the world changes around us.

It seems that the ideal way forward is for everyone to be willing to listen and learn from each other. The traditional, dare I say it, New Zealand school of thought could reasonably accept a little more compassion into its toolbox – and those with a complete aversion to killing could maybe examine the implications of inaction as well the horror of action.

Fortress conservation

In an effort to conserve biodiversity, areas of the planet are set aside for nature. Currently around 16 per cent of the planet's land area and inshore coast is in some way protected. This is, at face value, a good thing. Better still is the declaration from the Convention on Biological Diversity to push for 30 per cent to be protected by 2030 – the 30×30 target.

But … where is the most biodiverse land? Some 80 per cent of the world's remaining terrestrial biodiversity is to be found on land occupied by Indigenous people, who occupy around 25 per cent of the earth's land area and make up just 5 per cent of the world's population. And this makes many people in Indigenous communities worried, because what has happened before may well be repeated, all in the name of conservation.

Between 1990 and 2014 around 250,000 people have been evicted from their land in the name of conservation, reports *New Internationalist* magazine. Include the last century and the figure is close to 20 million people, all removed from their ancestral lands because other people thought they knew better. Even this pales beside the estimate from the human rights organisation, Survival International, which predicts that reaching the 30×30 targets will displace a further 300 million people.

Conservation has always had a dark underbelly, about which little is said. Yosemite was one of the first ever national parks, and one that created a model that was used around the world. It was granted its status in 1890 on the back of a long campaign of genocide. Between 1848 and 1900, the Indigenous population of California dropped by 90 per cent – as a result of disease, starvation and massacre. The remaining people in what became the park were driven off, and sporadic attempts to return to their homes were met with force. Now we know that the landscape was in part formed by the people, for example in their use of fire to protect oak trees that in turn provided them with the acorns that were central to their diet. But at the time, even the great John Muir dismissed the Indigenous Miwok, saying, 'They seemed to have no right place in the landscape, and I was glad to see them fading out of sight down the pass.'

In East Africa, the great national park of the Serengeti was created by restricting and removing the pastoralists who had used the land for generations. It is imagined, by founders and many visitors, as a land without a human history of any importance. A great threat to the integrity of this biodiverse wonder in Tanzania is poaching – the taking of wildlife by people who used to live with that wildlife, and who were forced from that land because someone else decided that they were more important.

'The main conclusion is that unless human population increase in areas surrounding protected areas is stopped, or even reversed, the future of conservation ... will be seriously compromised.' This was written in 2008 in a book called *Serengeti III: Human Impacts on Ecosystem Dynamics*. The authors do not go into detail as to what might be their final solution to the problem of these people.

The evictions do not end; there is an ongoing dispute in and around the stunning Ngorongoro crater. Here is a nature spectacle that fed my dreams of working with wildlife when Jane Goodall and her then husband, Hugo van Lawick, brought up their son, Grub (yes!) while studying the animals. Campaigning organisation Human Rights Watch have noted that since 2009 the Tanzanian government has used what it describes as 'a range of abusive tactics to displace around 150,000 people across the Ngorongoro district ... the government has restricted their access to important grazing areas and water sources.'

In 2017, the *Guardian* reported that security forces from the Tanzanian government had destroyed 185 Maasai homes in Loliondo, about 100 miles (160km) north of the Ngorongoro conservation area, in what human rights groups called a forced act of eviction.

In what is now the Democratic Republic of Congo, the Kahuzi-Biega National Park was established in 1970, and as it expanded it took over land lived on by the Batwa, who were forced out. Villages were burned, and while initially there was no direct killing, mortality brought on by the poverty caused by the evictions increased dramatically. Over time the park became a protected tourist destination and the authorities used an armed 'Rapid Intervention Unit' to enforce the displacement of the Batwa. When an attempt was made in 2018 by the Batwa to re-establish themselves in the park they were met with violence. At least 20 were killed, women raped, homes burned.

This style of management is known as 'fortress conservation' – sealing off national parks from the people who used to live there. More than 50 per cent of all nature-protected areas worldwide are to be found in the traditional territories of Indigenous people. At some point maybe it would be worth asking why the wildlife is still there, but not in the 'civilised' countries who are now so desperate to see nature protected? Could it be that the lifestyles of the hyper-consumptive developed world might just possibly be the problem? Not the Indigenous people who have been living with that land for generations? It is almost as if the authorities dictating where should be protected are still stuck in the racist mindset that allowed the Australian interior to be dismissed as terra nullius – nobody's land.

These beautiful places, rich in nature, are a rich person's playground, often built on the blood and bones of other, equally real, valid, sentient, feeling, loving, toiling and suffering people.

CHAPTER 11

Ethically consistent conservation – a manifesto

It won't have escaped your notice that there has been a lot of killing in this book. I wondered whether over time and exposure I would find it easier, but no, not in the slightest. I had also hoped that in the process of researching and writing this I would have found the absolute answer, the solution to the conflicts within wildlife conservation. Again, no.

But what I have found is that the more you dig deep into these stories, the more you realise how much more you need to know about nature and its interactions before you can make a decision. Ecology has always tended to be sidelined in the sciences – I know there are some who argue that the only true science is physics and the rest are simply cookery or stamp collecting, but leaving the blinkers of foolishness to one side, we can see that the great complexity of the wriggling web of life needs more attention.

Decisions about what to do are often taken by people with little grasp of the finer details. Yes, they may be advised by the wise, but that final decision will often be made without personal insight. There are few cabinet rooms in the governments of the world where we would so readily accept this sort of absence of understanding in the field of, say, economics. Yet economics is just a small subset of the ecosystem. Without a healthy and functioning ecosystem there is no functioning economy. We need a massive overhaul in priorities that is probably beyond the scope of this book.

Trying to pull all of the strings together to make some overarching sense has been taxing. While clarifying a detail from the work of mink man Tony Martin I mentioned how overwhelmed I was feeling. His response: 'Overwhelmed is good; I'd be worried if you weren't, because it would suggest that you're not taking it seriously!'

It is quite reassuring to have the person who managed to undertake the most dramatic and demanding of culls I have encountered accept that my head should be spinning.

So what have I learned? Have I become any wiser to this world? There are simple messages that make perfect sense – for example the call from the compassionate conservationists to 'Do no harm'. I love that – who wouldn't? But it is just too easy. Yes, it will sit well on a t-shirt, but it misses out too much to be really valuable. Maybe it could be developed into something like: 'Do no harm, which means we are going to need to build a time machine to travel back to stop us making a mess in the first place.'

If I really thought we could get this down to simple messages, then they would be as follows. Firstly, don't do it – don't go shipping animals and plants around the planet as it will only cause trouble. And secondly, protect what you already have. So often the trouble has started because the new arrival has found a niche – and while there are natural niches that can be exploited, there are also the ones we make by degrading what is already there. If our ecosystem is robust, diverse and abundant, then incoming species will find it harder to gain a foothold. But if we have hammered natural diversity and abundance into the ground, well, we have laid the table for our guests and must take some of that responsibility.

Sadly, none of that is going to resolve the troubles we have now. We, *Homo occisor*, were always going to make a mess – we are quite simply the most effective invasive species there ever has been. But that does not help me navigate a path through to something closer to a manifesto. I find there is a Goldilocks level of thought required to do this – we cannot act without thought yet we also risk thinking ourselves into stasis.

This is why a simple message can have appeal. When I asked Rachel King, campaigner at Animal Aid, for their stance, she was clear cut:

'Animal Aid has been campaigning peacefully since 1977 to end animal abuse and exploitation and to promote cruelty-free living in the UK. We believe that it is morally wrong to dictate which species can live (and where) and to demonise and kill others in the name of conservation. Culling can cause tremendous suffering and rarely offers a long-term solution, as other animals of the same species will often take the place of their culled conspecifics very quickly.

'Wildlife in Britain is rapidly declining due to destruction caused by humans, including urbanisation, agriculture, pollution and climate change. The conservation priority should be making sure wild spaces are protected and giving nature the time and space to thrive on its own, without unnecessary human interference. The species that we prefer to succeed may not, but we should let that happen rather than deciding to micromanage, control and inflict suffering on the species we don't like.'

They have embraced the same perspective as Marc Bekoff, and while I can appreciate why, I am stuck with the knowledge of what will happen if we simply do not intervene. The fact is that inaction can cause more death than action.

On Earth Day 2023 I gave a talk in Marston, Oxford. I had been booked months beforehand so had to miss out on the fun and games in London, where my wife was filming the protests of Extinction Rebellion and hanging out with a bunch of my lovely friends. I was a little frustrated at my prior commitment, but I always keep them; the importance of even small events can be hard to predict. Sometimes nothing, sometimes you sow a seed in the mind of a young person who goes on to grow great ideas into action. Or I get good cake.

This was a good talk – I had a full hall of people (though on reflection, there was no cake) and lots of conversations afterwards as people bought a few books. As I was beginning to pack up someone came and introduced themselves to me.

His name was Andrew Gosler – I had an inkling that I had heard the name, but could not quite put my finger on it. He is one of those people who carries his CV very lightly, and it was only when I got home and dug a little more that I found out he was a professor of ethno-ornithology at the University of Oxford, among many other intimidating titles.

We chatted about birds and hedgehogs and his work and mine. I was fascinated by the religious aspects of his life. He is now a minister in the Church of England, having transitioned from a life in the Jewish tradition. I did not at the time get to boast that I too was ordained – though my process was rather less onerous: a drunken night on the internet and a little bit of form filling means that I can officiate at weddings.

It was when we started chatting about this, the book I was working on, that he mentioned something really distracting. 'Oh, so you are writing about Hegelian Dialectics,' he said. I had to get him to repeat it so I could write that down. I had absolutely no idea what he meant, but I loved the idea that he could see what I was doing and give it a name.

When I got home I did some more reading. I was a little nervous that I might be wasting time heading down a philosophical rabbit hole, but was also quite intrigued.

Georg Wilhelm Friedrich Hegel was a German philosopher working in the nineteenth century, and he developed the dialectal ideas of Plato, which revolve around the interrogation of an idea in a process of dialogue. Hegel's spin on this was to take it on from a simple back-and-forth, and to develop it into a method of thesis, antithesis, synthesis. You put forward an idea, an argument, which generates a counterargument, which is the antithesis – and then rather than just returning to the argument you look to build on the merits of both cases to build the synthesis. This in turn may generate another antithesis, and on you go, hopefully refining the original idea into something that can hold water.

Well, if I had known that I was exploring Hegelian Dialectics I might never have had the courage to start – it seems so rarefied, but actually, on examination, quite valid. It really appears to equate to the scientific discipline: you see something, you have an idea about why it happens, you develop a hypothesis as to why it might happen, and then you test that hypothesis through experimentation, analyse the results and draw a conclusion.

To engage in this sort of conversation we need to talk to people outside our own little bubbles – it is all too easy to be reassured of the correctness of your position from within a space where no competing voices can be heard. This is what I set out to try and do – meeting people who have killed for conservation. Some were, on paper, easier to like than others – I found myself more excited at the prospect of meeting those looking to protect birds on Lundy and voles in Orkney than those who work to provide entertainment for armed nature rambles.

I can see how this happens. Too often the debates around killing for conservation get framed in the sort of simplistic binary I was at risk of collapsing into. I wonder how much of that is due to social media? I am beginning to realise how insidious this means of communication has become, with its capacity to disenfranchise moderate voices. The problem for anyone trying to work through these complex issues with honesty and integrity is that outrage generates attention, and attention is currency.

I think my aversion to conflict has helped! If I had started some of the trickier conversations at the point of disagreement, we would have got nowhere. But I instinctively dived into the areas we shared. Mike Swan might have been a gamekeeper but we shared a deep love of nature. It still makes me smile to realise the first bump I had with Jonathan Reynolds was about who recorded the better Sibelius *Fifth Symphony*, rather than his refinements to trapping technology. And at no point was I in any way burdened by whether or not I was engaging in the Hegelian method.

Strangely, when I look at the prejudice bundle I presented at the start of the book, it was with Wayne Linklater that I found some of the stickier territory. Not sticky to the point of a problem, but I found it hard to accept his laissez faire attitude to saving particular species for the sake of biodiversity. But, he was challenging – he asked questions that most people would never go near and he left me with more to think about. I am sure he is right and the Predator Free 2050 campaign will fail to meet its goals. And I can imagine those rats, in years to come, being protected as accepted members of the Kiwi community.

The only solution in New Zealand is going to be absolute protection of some areas – there is no way that there will be financial, or, as Wayne reminds us, social capital, to keep control going forever when eradication fails. And the only way to create those protected areas is going to be through killing. I fear it will be inevitable.

Marc Bekoff reminds us that the radical vision is to consider not killing to be an option. There was so much in what he said that I found attractive, and I really do hope we get to enjoy a glass of peaty malt whisky together one day. But when I look at the immediacy of the problems with hedgehogs, squirrels and rats, mink and stoats, there are so many good reasons to try and reset some of the balance that we have lost. Is that playing god? Do we have a right to intervene? Or possibly, would the irresponsible path be to not act?

This is, I believe, rather straightforward. We have, as a species, already intervened - and that presents us with a choice. To leave our mess for 'nature' to clear up over time – causing the extinction of species along the way, and possibly irreversible changes to wider ecosystems – or to step in and step up. Marc's argument about the timescales on which we judge these things is well made, but also, despite its superficial humanness, very unhuman. We are not built like that – would we stand by as a fire struck the Louvre and decide that we should just let this happen, as more art will appear in the years to come?

That then gives us the next step. If were are playing god, how do we do it? Are the non-human animals we are working with moral entities? Should they be given moral consideration? Do they count?

I think you would have to be an inherently cruel person to think that the wildlife we share this planet with is not worthy of moral consideration. This is not to leap into the arguments I used to have at school when talking about animal rights – the adults would always retort with something along the lines of: 'If they had rights they would need responsibilities and I don't see any of them paying tax'. Fortunately there tends to be a slightly more informed and nuanced debate these days (I say that while desperately ignoring some of the less informed and utterly un-nuanced debate that takes place on social media).

As we saw with the shifting of goalposts when animals were seen to use tools, trying to lay claim to the exceptional nature of our species is tricky – especially when you consider that a hard line would disenfranchise infants, those suffering from dementia, and the very drunk.

And as the former deputy chief vet of the United Kingdom, Alick Simmons, so clearly states, the animals we are seeking to manage all have the capacity to experience pain and fear. So there is no way that we humans can return to the world of Aristotle and Descartes – or perhaps we need to ensure that we work hard on those who still think of non-human animals as automata.

The people I met along the way who have made it their life's work to kill, get no joy from the killing. Whether it is the rats on Scilly or Lundy, the stoats of Orkney, the mink of East Anglia, or the grey squirrels of Anglesey, none of these are killed in rage, or, for that matter, in glee. They are killed with resignation, and with the sense that it is an acceptance of our failure to solve the problems, either at source, or in a more creative way.

Though the feelings of the people doing the killing – should those be important? This is a question I have put to students during lectures. Do they think it makes a difference in these settings if the person undertaking the task gets a kick out of it? I have been surprised at the lack of interest and thought that many were willing to give the question, but to me it seems central. Let's take the act of killing an elephant, for example. This might be done because habitat fragmentation has caused more elephants to be confined in an area that is too small to support that number.

So the decision has been made that the elephant has to be killed. Do you give the job to a professional who does it because it is an important job? Or do you hand the task over to someone who gets their kicks from killing? While the latter may bring in money, it also brings with it disgust.

The images of hunters leering over the corpses of beautiful animals is obscene. It generates disgust – a very powerful reaction, and one that is non-proportional, in that once it has been triggered, subtlety and nuance evaporate.

Does the killing need to happen? That is perhaps the most fundamental question, and it must be central to any process of decision-making. It is evident from the experts on the ground that there have been and will continue to be extinctions of species caused by invasive non-native species.

If you feel – when all other avenues have been exhausted, and surely that must be the way – that killing is always and only the last resort, then you have to take responsibility for the consequences, for the death of the condemned. And likewise, if for you the evidence presented leaves you feeling that killing is too great a cost to pay, then you have to accept the consequences for the loss of those individuals and species affected by the presence of the non-native species.

But if you can accept that in certain circumstances, killing for conservation may be valid, then I think it is important to look at ways of ensuring it is done in the best way possible. And if you have got this far, this is the moment you have stepped out of the binary and into the grey. Marc said how the grey is 'always leaning towards killing'. This seemed wise at the time but on reflection is simply a statement of the obvious. If the position held is absolute, where the only acceptable answer is zero, adding anything to that, any subtlety, inevitably leads to something – from black and white come grey.

In writing that, I am clearly nailing my colours to the mast of utilitarianism over deontological, or normative, thought. When I first read Peter Singer's discussions of animal rights, I assumed he was advocating the normative approach – thou shalt not kill. But as I said, that was as a child, and one who was clearly out of his depth. Now I see how Singer actually embraces the utilitarian view, the minimisation of suffering. I wonder how many other people with similarly strongly held views are actually hiding their realisation of a more nuanced world view under the baggage of prejudice?

Problems arise, of course, from accepting a more subtle debate. If you are an organisation that relies on absolute and polarised messaging for fundraising, then the space for nuance can be limited. And obviously the preferred method of transmission, through social media, relies entirely on the engine of outrage, thus further eroding space for thought.

This happens on both sides of the debate – there is no way that Chris Packham would have been in receipt of quite the surge of hate and death threats if it were not for the very simplistic messaging of his opponents. And it is intriguing to discover that the farthest wings of the animal rights world will sometimes spend as much time attacking moderates within their own movement as they will the actual animal abusers. This is what creates fear among the NGOs that have funds to raise.

Social scientists have been able to offer a really useful insight into the way that the polarisation of debate can occur, and how it can be best overcome. For example, Isla Hodgson looked into the arguments around grouse hunting in Scotland. An important result from this was that while most of what is seen – and heard – of the debate is between the two extreme ends of it, the vast majority of people sit within the centre ground. The extremes outdo each other in outrage and create further polarisation. This 'turns off' the middle ground, where lies more chance to compromise. The end result is that less gets achieved.

Even the most hardworking and dedicated animal rights supporting workers at one of the most important wildlife rescue centres in the United Kingdom admit that they have to kill some animals that pose a risk to their ability to operate. They do not like killing rodents but if they did not, then the authorities would shut them down. And shutting the hospital would lead to the death and suffering of many other animals.

From this we can see that the utilitarian arguments put forward by Peter Singer really do require moderate voices to be creating the change. That is not to say that there is no value in having high standards. There is no point asking for a 2 per cent pay rise from your boss when they offer you 1 per cent – you ask for 10 per cent and hope to get to 5 per cent, you haggle. So by starting from a point of no killing, but by being willing to engage in the debate, it is possible you will help create a better outcome.

Or as the brilliant songwriter Robb Johnson sings, 'Be reasonable, demand the impossible now.'

When we confront these issues we really do need to take a brave step back and do, as I have attempted to do, examine our motivations, our prejudice. In its simplest form, this is head-versus-heart territory. But that does not present us with a clear path.

Well, a path was worked on a few years ago when 20 experts, hosted by the University of British Columbia and the BC Society for the Prevention of Cruelty to Animals, met for two days to explore the ethical and evidence-based approaches to managing such conflicts. The result was a peer-reviewed publication called 'International consensus principles for ethical wildlife control' which developed a seven-point approach.

Efforts to control wildlife should:

1. begin wherever possible by altering the human practices that cause human–wildlife conflict and by developing a culture of coexistence;
2. be justified by evidence that significant harms are being caused to people, property, livelihoods, ecosystems, and/or other animals;
3. have measurable outcome-based objectives that are clear, achievable, monitored, and adaptive;
4. predictably minimise animal welfare harms to the fewest number of animals;
5. be informed by community values as well as scientific, technical, and practical information;
6. be integrated into plans for systematic long-term management;
7. be based on the specifics of the situation rather than negative labels (pest, overabundant) applied to the target species.

The lead author on this was Dr Sara Dubois, Chief Scientific Officer of the British Columbia Society for the Prevention of Cruelty to Animals. I was lucky enough to book her into a Zoom call, and while this technological innovation is a wonder, I would have rather been with her in person, especially when she indicated quite where she was calling from – Vancouver island.

Sara is fascinating. She is a rigorous scientist but with a very open heart; managing to be sensitive to the world without drifting into 'woo', that sort of vague mysticism that really grinds my gears. 'A frustration I have,' she said, 'is the lack of reverence to individual life.'

It was such a simple statement but it pulled me up. I have mentioned the problems that come with language – 'othering' to ease in more dramatic action – but I don't think I had fully appreciated the simple notion quite as clearly as now. That every time wonderfully thoughtful ecologists stride out into the wild to fix a problem, it is not pests that are being killed, nor is it rats, stoats, mink that are being killed, it is individuals of those species.

That is not to say it should not be done. But it should not be done without full awareness of the fact – and it should not be done without the time to consider the uncomfortable nature of the decision.

'Biologists go into their work with a good heart,' Sara said, and this is where her life began too. 'But over time their capacity to see individual animal suffering is scrubbed clean.'

One of the contentious issues in her part of the world is the killing of wolves to help caribou. 'There have been repeated plans as to how to increase caribou numbers,' she explained. 'And in doing so they have failed to meet most of the principles for ethical wildlife control. Let's take the first in that list – changing human behaviour. If this had been done then they would have prioritised habitat protection and reduced the impacts of development on caribou. There is a serious risk of losing caribou populations, so this merited serious intervention. But because effort was put into restoring caribou numbers, not their habitat, this meant that wolves, as their natural predator, would be presented with a higher concentration of caribou and easier access from logging roads, becoming themselves the target of control when they start to hunt their prey.

'The fourth on the list – minimising harm; shooting from helicopters carries a high risk of injury and is therefore inhumane,' she continued. 'And then there is the use of Judas wolves. Again, so inhumane.'

I was unfamiliar with that term so Sara explained, and it is now in the filing cabinet in my brain where I store the material I really wish I did not know.

'One of the methods used to kill wolves is when they capture an individual and attach a GPS collar to it and then wait for this, the Judas wolf, to lead them to the rest of the pack, who are shot from a helicopter. But – and this is the bit that takes it into a different league of inhumanity – the collared wolf is left alive, and in time leads the hunters to a new pack, who are then shot. And on it goes.'

'Combine this with a lack of systematic, long-term management plans, other than killing as many wolves as possible, and the continued failure to solve cumulative impacts like habitat loss and degradation, forest fires, climate change, caribou health, disease, and reproduction. Well, this is why we wrote the principles, so that we could learn. And this is a mistake we could all learn from.'

In fact Sara is equally passionate about mistakes.

'It takes courage to write about your mistakes – you might risk funding and publication opportunities. But we learn from our mistakes and others do too. Yes, if we follow the ethical principles we should reduce the chance of something going wrong, but if things do not go according to plan, we need to acknowledge that so that we can improve our work. Because when we do this work better, fewer animals suffer and die.'

And there we have both the scientific and the Hegelian dialects coming together – refinement of the approach until, maybe, we reach a point where the killing can cease.

I was interested that Sara had a list of seven key principles; Tony Martin also presented a list of six key questions in the chapter on mink. They were more practically minded and focused on the species in mind. The RSPB, as I mentioned earlier in the book, has a shorter list with just four questions you can summarise as: is it serious; are there no non-lethal measures; will killing address the problem; will there be wider conservation consequences. And the advocates of Compassionate Conservation have their quartet: 'Do no harm; Individuals matter; Inclusivity; Peaceful coexistence.'

The closest I have yet found to a reasonable pathway through the issue of killing for conservation does come from the work of Sara.

Lacking in the guidance from the RSPB and Tony Martin is any references to cruelty. Sara does talk about minimising welfare harms, but I think that cruelty is something that has to be considered and addressed. The suffering of the rats on South Georgia following consumption of brodifacoum was not part of Tony's equation.

Laboratory studies have shown that it may take four days for a rat to start showing signs of poisoning from the point of consumption. This is crucial; it is important that the bait is not associated with illness, so that it keeps being consumed. Typically it takes a further four days to die. I have read the literature describing the suffering rats go through – it is deeply disturbing and in one paper was summarised as 'anticoagulants in general were ranked amongst those producing the most severe and prolonged poor animal welfare.'

In the same way that someone utterly opposed to any killing must accept responsibility for the death that their decision would bring, so people who use lethal methods must accept their part in the incredible suffering their work causes. Which is why Tony believes trapping mink on remote islands would be impossible to conduct humanely, as inevitably, animals would be left for extended periods, possibly to the point of starving to death. Whether that is a worse fate than poisoning, though, is a moot point.

The progress towards agreement on issues like this is so challenged by the binary nature of so-called debate, killing or no killing. Whether it is Brexit or trans rights, really subtle and complex issues get reduced into soundbites and abuse. I love social media – I use it to promote my work, to campaign, to keep in touch – but there is also a dark side that was perhaps weaponised most effectively by the former testiculator in chief of the US. Trump and his ilk thrive on division.

Value on social media is measured in attention. Attention is most easily found through outrage. The best way to get outrage is to be controversial and angry – and while that might start as a means to an end, it inevitably results in those sentiments sinking beneath the skin. Wear the mask long enough and when you take it off, there is no perceivable difference.

A psychological study at Yale University showed clearly that social media has the ability to reinforce positions of social outrage. It is fascinating to see how this unfolds, and, when spoken, so obvious. But while it is blatant it is also largely ignored. Participants in the study at Yale who received 'likes' and 'retweets' when expressing outrage were more likely to express outrage in later contributions.

Outrage is not all bad – it can be a strong force for good, encouraging cooperation, generating change. But it also helps to harass minority groups, spread disinformation and create polarisation.

Lead researcher Molly Crockett, now associate professor of psychology at Princeton, said, 'Amplification of moral outrage is a clear consequence of social media's business model.' Interestingly, amplification was not consistent across different groups. The more politically extreme networks obviously expressed more outrage than members of more politically moderate networks.

'Our studies find that people with politically moderate friends and followers are more sensitive to social feedback that reinforces their outrage expressions,' said Crockett. 'This suggests a mechanism for how moderate groups can become politically radicalised over time – the rewards of social media create positive feedback loops that exacerbate outrage.'

Does this mean that the extreme views expressed around killing for conservation can be dismissed as fabrications of the platforms on which they appear? I don't think so. They are there and they are deep. In fact I found an issue that created such tension that I ended up deciding it was not worth diving into. Trophy hunting.

Could the money earned from trophy hunting be useful in ensuring the conservation of species and habitats? The voices are strident and unyielding. So much so that there was no space for debate – entrenched ideological warfare with both sides lobbing tweets instead of mortars. Even starting to get involved, I could see that it was going to lead to any moderate voice being shouted at by either side – it was as Isla Hodgson revealed when talking about grouse hunting.

I had imagined that there would be some potential for shared thinking, but ended up realising that so contentious was the subject that it would end up swamping the book. Which is frustrating as I can see where the problems lie. In fact it was a comment from a friend, Sian Sullivan, who also happens to be a Professor of Environment and Culture and at Bath Spa University, who gave me the insight by saying, 'Fundamentally, I believe that ethics should precede science.'

Which spins me back into the more normative (deontological) way of thinking – some things are wrong.

One of my possibly disagreeable prejudices comes in the form of innate distrust of politicians, unless they have done something to challenge my bigotry. There are those that have found a place in my heart; having had the pleasure of meeting both Tony Benn and Caroline Lucas on a few occasions I can attest to their decency. But never had I found myself warming to an MP from the Conservative Party until a conversation with the then minister, Rory Stewart, in 2015.

His office had called to arrange this, as he was due to present an answer to a question about hedgehogs in the House at the request of the British Hedgehog Preservation Society that we identify the hedgehog as England's national species. I talked with Rory for around 45 minutes and was taken by his obvious intelligence and quite forensic questioning.

He gave his speech the next evening, and while I was obviously disappointed that he spoke against the idea of having a hedgehog icon, he did present what has been widely quoted as one of the most brilliant speeches in the House of Commons for many years. Thirteen minutes, dedicated to hedgehogs, the first time they had received such attention since 1566, and it includes my one and only reference in Hansard, when he said, 'It [the hedgehog] has been preserved thanks to eccentrics such as my hon. Friend and, perhaps most famously of all, Hugh Warwick, the great inspiration behind the British Hedgehog Preservation Society. He has written no fewer than three books on the hedgehog, and he talks very movingly about staring into the eyes of a hedgehog and getting a sense of its wildness from its gaze.'

And while I bask in the glory of this inclusion, I have to stress that the BHPS was very much working on hedgehogs before I got involved.

The reason I bring this up is actually because of some other prickly work Rory has got himself involved with, when he presented a fascinating three-part series on BBC Radio 4 called *A Long History of Argument*. Through this he explains how argument is so important – how it developed the ways we answer the deepest questions of philosophy, science and the law. How it was the foundation for our democracy, and how truth can be found through the process of argument. But his time as a member of parliament showed this to be somewhat wishful thinking, and now arguments are used to provoke division and hide truth.

In particular he described how he saw a shift in political discourse – how he had always seen it as a bell curve, what is known as the normal distribution curve. For example, imagine measuring the height of everyone in your country and plotting it on a graph: most people will be in the main body of a bell-shaped curve, with a few outliers at either end. When we study ecological data we tend to look at the main body containing 95 per cent of the information, with just 5 per cent being outliers. Anyway, imagine that in what Rory sees as the good old days, most people sat within that 95 per cent of the bell, when it came to debate around complex issues.

However, the change since 2014, which has been born off the back of outrage-fertilised social media, is for that curve to become inverted. Most political discourse now takes place on the extremes, with what was the central ground either being silenced – for fear of attack from either extreme – or, more worryingly, pulled to the extremes, echoing the work from Molly Crockett at Princeton.

As Rory points out, this is not conducive to the creation of a better world. Though he did inadvertently give another argument for the great value of the hedgehog when interviewed about his radio series. He was asked why the hedgehog speech was the one that everyone remembers. He responded, 'It is because there is something magically appealing about hedgehogs. And in a political world so fractious and binary, it is a subject on which everyone can soften and converse and be human.'

I have said before and I will say it again – the hedgehog is massively underrated.

Before we can make any progress, though, we need more than just good science to explain the ecological ramifications of the actions we choose to take. We need good communications. I think that Rory Stewart's analysis of the polarisation of debate is astute, but we need to go beyond just understanding what is wrong and look to how things can be turned right. And for that, it is back to the thinkers, the philosophers – and in particular, Bertrand Russell who revealed himself to be remarkably prescient in an interview from 1959, when he talked about love being wise and hatred being foolish. In particular it was his awareness of the need to learn tolerance – because even then he was seeing a more interconnected world.

The cleverness of youth may be replaced by the wisdom of age – the desire to win each fight exchanged for a view of the big picture that might just be more complicated than we thought. I hope we can become more and more like hedgehogs, leaving our skittish inner foxes behind.

In a 1934 essay Russell also gave this wonderful instruction: 'Perhaps, instead of teaching manners, parents should teach the statistical probability that the person you are speaking to is just as good as you are. It is difficult to believe this; very few of us do, in our instincts, believe it. One's own ego seems so incomparably more sensitive, more perceptive, wiser and more profound than other people's. Yet there must be very few of whom this is true, and it is not likely that oneself is one of those few. There is nothing like viewing oneself statistically as a means both to good manners and to good morals.'

Being willing to see ourselves more honestly, more as we are than we might think we are, is at the heart of a new approach to the environment. I was doing a talk at a festival straight after Extinction Rebellion co-founder Rupert Read. He had a daytime job as professor of philosophy at University of East Anglia and has spent a very long time thinking about how to create the population-level change that is required to meet the goals of the environmental campaigners. His talk was about the rather clumsily named 'moderate flank' – the sorts of people who might not normally be associated with radical action, but actually share many of the same values.

Likewise a deeply wise friend of mine George Marshall – a climate communication guru if ever there was one – has for years been pushing the line that we need to talk to different people. He argued that the worst thing that happened to climate change was that it was picked up by the environmentalists before everyone else – that it became about polar bears rather than jobs.

'I have travelled through the most conservative parts of the USA and sat with people who hold diametrically opposed views on the issues of climate change,' he said over a pint of beer. I wanted to know how his talks went down, but had missed the point. 'I was there to listen, and came away realising that these are, mostly, lovely normal people who have been fed very different stories. Stories that are immediate, rather than the long-term sagas of the environmentalists. Their stories are grounded in their local landscape, community and values, rather than the globalised moralistic preaching of environmentalists.'

Back in the United Kingdom, George realised that the groups that had the greatest potential to create active change were not the local chapters of Extinction Rebellion, but the local churches or conservative clubs. 'If you want something to get done, properly and on time, I would tend to rely on the people who have managed to keep a church and community centre running for decades, who have organised fundraising cake sales and summer fetes. These are the people we need to share our stories with.'

Rather unexpectedly I find that I have been doing the same sort of thing for a very long time. As you might have guessed, I talk about hedgehogs, a lot. I get invited to the Women's Institute, University of the Third Age, gardening clubs and many others to share my passion for hedgehogs. I am driven by the very high standards of cake on offer, and the opportunity to talk about so much more. Because when I talk about helping hedgehogs I am also talking about the need to regenerate macro-invertebrates in the farmed landscape, which requires us to reconsider agricultural policy and dietary choices. We need to talk about transport infrastructure and the importance of the National Planning Policy Framework. And we also need to – in as light-hearted a manner as possible – bring up the idea that industrial capitalism is not a model that can sustain life for much longer, and wouldn't it be great if we could look for an alternative.

None of these groups would have booked me to talk about these subjects, but I have, in effect, created a 'trojan hedgehog' to snuffle my way inside these often highly conservative gatherings and then unleash ideas that are new to many.

Anyway, what Rupert Read, George Marshall and I are doing is trying to speak and listen outside the bubbles within which we can easily find confirmation and support. And so it is with the issues raised in this book. We can retreat into our bubbles, and shout insults at people in other bubbles. But nothing will change.

I find it interesting to look back at the version of me who started out on this book; I really hoped I would find evidence to take a robust stand, one way or the other, about killing for conservation. I realise that was in large part down to cowardice – if I could take a stand I would have allies, people who would shout for me in the face of criticism from the others. But I began to realise that this is actually part of the problem – setting out to try and find a simple solution.

Ecological problems are going to require ecological solutions which in turn will mean we have to think ecologically. We need to jettison the simplistic and the binary and be willing to think better.

I have been reading and listening for a long time now and I have found myself being continually distracted by loud, clever, and persistent voices, articulating what seems to be one side or the other of a deeply complex and subtle problem. It really feels like we are missing out on complex and subtle thoughts.

Complex and subtle thoughts get drowned out by the binary – echoing down from the walls of the wine glass; all noise, no wine! Extreme views are in part a product of, as we have seen, the means of dissemination. But they also take root in people's insecurities. The more harsh or undernourished a life has been, the more understandable insecurity there is, which allows for easy answers to difficult problems to find fertile ground.

When a non-native species arrives in a new environment it will only be able to settle down if there is space, if there is an ecological niche available. Often that niche is available because we have degraded the native environment, giving space for the new arrival to thrive. Additionally, as seen in New Zealand, when the home environment lacks diversity and abundance of native life, the impacts of a newcomer are all the more obvious.

The parallels are not too hard to see. If we do not look to our home ground to ensure health, diversity and abundance, we will be more open to both polarised thinking and invasive species. The answers are not quick and easy, they require years of investment in society and in nature. But the benefits will be rich.

'A mature person is one who does not think only in absolutes, who is able to be objective even when deeply stirred emotionally, who has learned that there is both good and bad in all people . . .'

The words of Eleanor Roosevelt make me hope that one day I can aspire to a bit more maturity.

But I do not despair – that is only for those who have seen the end of the story, and as we never can, hope is the only way forward. And what I hope for is that we recognise the danger of an empty wine glass, and we fill that void with thoughtful conversation, a willingness to brave the world outside our own bubbles, and above all, good wine.

Acknowledgements

This has not been an easy book to write. I have had to dive deep into unfamiliar waters – and I have had to confront and challenge my prejudices. Some of the bedrock of support came from the scarily intelligent philosophers I am lucky enough to have as friends. Angeliki Kerasidou, Amna Whiston and Paula Casal all carried out various levels of handholding.

I have spoken to so many people about this work – I find that in speaking through the ideas my thoughts often clarify. Thank you to all who have been subject to this. As ever, the orchard crew have been particularly supportive: George Monbiot, Rebecca Wrigley, Roman Krznaric, Kate Raworth, Caspar Henderson, Phil and Amanda Mann – and not forgetting the younger members – Siri, Cas, Henry, Isabella, Martha and Lara.

Wise friends are good to have – ones who will listen and are not afraid to comment: Susan Canney, George Marshall, Tom Moorhouse, Robin Ince, Chris Packham, Megan McCubbin, Amy-Jane Beer, Julian Hector, Dominic Woodfield, Nell Frizell, are among many who have suffered my moments of lost confidence, or just my simply being lost.

A huge thank you to those who let me into their lives and shared their stories and thoughts with me – Alick Simmons, Craig Shuttleworth, Rosie Ellis, Jaclyn Pearson, Sarah Sankey, Tony Martin, Jonathan Reynolds and Mike Swan gave me coffee, food, time, and lots to think about.

Given the limitations of time and money I was not able to travel the world to meet some of the amazing people I have had the pleasure to interview remotely – that is a shame as in all instances they just came across as so intelligent and generous – and challenging (in a good way). One day I hope to meet Katie-jo Luxton, Chris Jones, Wayne Linklater, Marc Bekoff, Monica Engel and Sara Dubois in person.

I did get to spend time with Mary Colwell, Ricarda Steinbrecher, Preen Singh Bindra and Andy Gosler – more of which is required.

I think that everyone should have a friend like Miriam Darlington – willing to dive into an adventure at short notice, being necessarily critical as well as kind and supportive.

The book is only now in your hands thanks to the work of James Macdonald Lockhart, my agent, who helped persuade Julie Bailey at Bloomsbury that this was a worthwhile enterprise. Along with her team – Amy Hodkin, Jessica Gray and Erin Brown – they have all been patient, and unfailingly understanding of the challenges a book like this presents.

Long suffering is an over used term by authors in their acknowledgements when referring to their family … but I completely feel it is legitimate to be used in this case! Thank you Zoe, Raine and Pip for being there. And now its Zoe's turn – need her book finished soon … and Zoe has reminded me that I ought to thank our lovely rescue hound, Ogli, even though he very much prefers her!

And thank you to Anne – my mother.

Family comes in many forms – my 'odd' children, Silva, Naia, Rosa and Jagusia – you are amazing and I can't wait to see what you become. I love the faith that has been placed in me by your parents – Chloe, Christian, Lisbeth, Oliver, Theo, Shannon and Kasia. Gaz and Jools, Raya and Tiger – for being another family for Raine, thank you.

Eylan Ezekiel – what a delight to have you and your family as friends and neighbours – and that is magnified by your gift to me of the title of this book. I had been struggling for so long. A strange impact of covid has been bringing our cul de sac together as a supportive group – what an ace bunch.

Along the way I have received great support from many people – particularly I would like to thank Jane Byam Shaw and Derek Gow for providing holiday spaces to let me and the family decompress; Gerry and Lucie – for letting me loose in their woods; Roz Kidman Cox for starting me on the writing path. And Simon – for music that heals.

Further reading

This is not a complete and exhaustive list of the important reading I did along the way and I did not approach writing the book with academic referencing in mind. These are just some ideas that will help you dig deeper into the world of this book.

Bekoff, M. (ed). *Ignoring Nature No More: The Case for Compassionate Conservation* (University of Chicago Press, 2013)

Colwell, M. *Beak, Tooth and Claw: Why We Must Live With Predators* (William Collins, 2022)

Edmonds, D. *Would You Kill the Fat Man?: The Trolley Problem and What Your Answer Tells Us about Right and Wrong* (Princeton University Press, 2015)

Gruen, L. *Ethics and Animals: An Introduction*, 2nd edition (Cambridge University Press, 2021)

King, C. *Immigrant Killers: Introduced predators and the conservation of birds in New Zealand* (Oxford University Press, 1984)

Lewis, S and Maslin, M. *The Human Planet: How We Created the Anthropocene* (Pelican Books, 2018)

Marris, E. *Wild Souls: What We Owe Animals in a Changing World* (Bloomsbury USA, 2023)

Simmons, A. *Treated Like Animals: Improving the Lives of the Creatures We Own, Eat and Use* (Pelagic Publishing, 2023)

Yalden, D. *The History of British Mammals* (T & AD Poyser, 2002)

Index